Wind Power Generation

The Power Generation Series

Paul Breeze—Coal-Fired Generation, ISBN 13: 9780128040065

Paul Breeze—Gas-Turbine Fired Generation, ISBN 13: 9780128040058

Paul Breeze—Solar Power Generation, ISBN 13: 9780128040041

Paul Breeze—Wind Power Generation, ISBN 13: 9780128040386

Paul Breeze—Fuel Cells, ISBN 13: To Come

Paul Breeze—Energy from Waste, ISBN 13: To Come

Paul Breeze—Nuclear Power, ISBN 13: To Come

Paul Breeze—Electricity Generation and the Environment, ISBN 13: To Come

Wind Power Generation

Paul Breeze

AMSTERDAM • BOSTON • HEIDELBERG • LONDON
NEW YORK • OXFORD • PARIS • SAN DIEGO
SAN FRANCISCO • SINGAPORE • SYDNEY • TOKYO

Academic Press is an imprint of Elsevier

Academic Press is an imprint of Elsevier
125, London Wall, EC2Y 5AS.
525 B Street, Suite 1800, San Diego, CA 92101-4495, USA
225 Wyman Street, Waltham, MA 02451, USA
The Boulevard, Langford Lane, Kidlington, Oxford OX5 1GB, UK

Notices
Knowledge and best practice in this field are constantly changing. As new research and experience broaden our understanding, changes in research methods or professional practices, may become necessary.

Practitioners and researchers must always rely on their own experience and knowledge in evaluating and using any information or methods described herein. In using such information or methods they should be mindful of their own safety and the safety of others, including parties for whom they have a professional responsibility.

To the fullest extent of the law, neither the Publisher nor the authors, contributors, or editors, assume any liability for any injury and/or damage to persons or property as a matter of products liability, negligence or otherwise, or from any use or operation of any methods, products, instructions, or ideas contained in the material herein.

ISBN: 978-0-12-804038-6

British Library Cataloguing-in-Publication Data
A catalogue record for this book is available from the British Library

Library of Congress Cataloging-in-Publication Data
A catalog record for this book is available from the Library of Congress

For Information on all Academic Press publications
visit our website at http://store.elsevier.com/

Working together
to grow libraries in
developing countries

www.elsevier.com • www.bookaid.org

CONTENTS

CHAPTER *1*

An Introduction to Wind Power

Wind power is one of three major renewable energy resources, alongside solar power and hydropower, that are being exploited on a large scale for global power generation. As an energy resource, wind is widely distributed and it is capable of providing power in most parts of the world but it is both intermittent and unpredictable, making it difficult to rely solely on wind for electrical power. When used in conjunction with other forms of generation however, or in combination with energy storage, wind can make a valuable contribution to the global energy balance.

According to the International Energy Agency, the contribution of wind energy to total electricity generation, worldwide, rose from 0.2% in 2000 to 2.3% in 2012,[1] making it the second most important renewable generation source after hydropower. This is still a relatively small proportion of global output, reflecting the fact that wind power has only become a mature technology in the twenty-first century. In contrast, hydropower, the best established of the renewable technologies, contributed 16.2% of global electricity production in 2012.

The modern development of a commercial wind power industry started in the 1970s in the United States before spreading to Europe where Denmark and Germany became significant champions of the technology. More recently the focus of growth has shifted to Asia and particularly China where there has been massive development of wind power during the past decade. The growth of the wind industry has followed the development of the technology with major wind companies in the United States, Europe, and China. This has led to a level of global competition within the wind industry. Even so, many contracts are still won by local companies and wind turbines cannot yet be considered a global commodity.

---

[1]World Energy Outlook 2014, International Energy Agency.

Wind Power Generation. DOI: http://dx.doi.org/10.1016/B978-0-12-804038-6.00001-3

Most wind power plants in use today are located on land. However a small but increasing number are being built offshore, in coastal waters. Building offshore is much more difficult than installing wind turbines on land because a foundation for the turbine has to be created in the sea bed. Against this, offshore wind farms generally raise fewer objections than onshore developments, so gaining permission to build can be easier. In addition the wind regime offshore is often better than that onshore so an offshore wind farm can generate more electricity than a similarly sized plant onshore. In the future offshore wind turbines may be mounted on floating platforms, making deployment much easier.

Offshore wind turbines are generally larger than those installed onshore. Modern wind turbines are extremely large machines and transporting components, such as the mast and blades, to an onshore site can be difficult as all these parts must be transported by road. Transportation can restrict the absolute size of onshore wind turbines, although the difficulty may be overcome in the future with new, modular designs. There are fewer restrictions offshore where components are carried to the site by ship, so building wind farms offshore allows bigger wind turbines to be installed.

The experimental use of wind energy to generate electricity began in the nineteenth century but it was the oil crisis of the 1970s that finally stimulated the commercial development of wind power. Early machines were produced to a range of different designs and most were both small and primitive by twenty-first century standards. Wind farms began to appear in the 1980s, particularly in California where farms comprising hundreds of generating units were installed. Interest in wind power waned during the late 1980s and early 1990s but then the development of advanced, megawatt sized turbines later that decade, combined with the recognition of global warming and its causes, encouraged development to accelerate once more. It is machines of 1 MW and greater in capacity that now dominate the global industry. Even so, there remains a market for small wind turbines, particularly in countries like the United States with a history of off-grid generation.

The intermittent nature of wind means that wind power must always be supported by another source of electricity. On an isolated system this might be some type of energy storage but with grid-connected turbines and wind farms it most usually takes the form of a fast-acting grid backup, often based on gas turbines. Managing wind power on a modern grid also requires sophisticated forecasting techniques so that these alternative sources can be made ready for periods when the wind does not blow. This makes the balancing of a grid with large volumes of wind power much more difficult than one supplied by conventional generation sources alone.

Wind power is considered to be economically competitive with alternative sources of generation in situations where the cost of grid electricity is high but the technology has not yet achieved across-the-board parity with the main fossil fuel-based generation technologies such as coal and natural gas. However if the environmental cost of fossil fuel combustion is added into the equation, usually in the form of a cost for each tonne of carbon or carbon dioxide released into the atmosphere, then onshore wind power can attain parity with virtually all the alternative forms of generation. The electricity from offshore wind farms still remains more expensive than these alternatives.

THE HISTORY OF WIND POWER

The Greek engineer Hero of Alexandria is credited with devising, in the first century, the earliest known machine that was powered by a wind turbine. In his case the machine being driven by a wind was a wind-powered organ although there is no way of discovering if the machine described in his writings was actually built. Wind machines were developed for more conventional applications such as grinding grain or pumping water in Iran and Afghanistan around the ninth century, possibly earlier. These were vertical axis windmills in which a vertical shaft was attached to a series of vanes or sails covered in cloth. From this region, the use of windmills of this type spread across Asia to India and China.

The more familiar horizontal axis windmills were first recorded in north-western Europe in the twelfth century where they were used for milling and for pumping water. The configuration of these windmills is significantly different to those found in Asia and the design is believed to have been developed independently of the earlier type. In Europe, wind power offered an alternative to the widely used water mills with the attraction that they could be built anywhere, not just along water courses. They were used in many parts of northwest Europe and became particularly popular in the Netherlands for pumping water in low-lying areas. Many of these old windmills are still operable.

Wind power became popular in the United States during the nineteenth century where small, multivaned wind pumps were used as a means of pumping water to supply irrigation for crops. Although the early versions of these mills were made from wood, they were later made from iron or steel and many of these wind pumps can also still be found across rural America, particularly in the Midwest. It has been estimated that between 1850 and 1970 more than 6 million windmills were erected across the United States.

The discovery of electricity, and the subsequent development of the dynamo which could convert rotary motion into a direct current, soon led to the adaptation of windmills for electricity generation. The first recorded instances were both in 1887.

In Scotland, in July 1887, James Blyth built a relatively primitive cloth-sailed vertical-axis windmill with a 10 m diameter rotor that produced electricity to charge accumulators (early batteries) and supply lighting for a cottage. Over the winter of the same year Charles F. Brush—the designer of the Brush dynamo—designed and built a horizontal axis wind turbine in Cleveland, Ohio, with a 17 m rotor and 144 blades that drove a 12 kW dynamo. This turbine was used to charge accumulators too, as well as to power arc lamps and electric motors. The machine was superseded in 1900 by centrally supplied electrical power but remained operational until 1908 or 1909. Meanwhile in the 1890s the Danish scientist Poul la Cour began developing wind turbines that could be used to bring electrical power to rural areas of Denmark.

During the 1920s French engineer George Darrieus invented the vertical axis wind turbine, often called an egg-beater because of its curved blades, that bears his name today. Meanwhile Russian engineers built a 100 kW horizontal axis machine at Yalta in 1931. The first megawatt scale turbine (the capacity was 1.25 MW) was erected at Castleton, VT, USA, in 1941.

Sporadic wind turbine development continued until the 1970s when, spurred by a major oil crisis in the Middle East, the US government, through NASA, began to fund the development of a new generation of wind turbines. This led to the evolution of many of the ideas for construction and control used in modern turbines including composite blades and variable speed generators. From this program sprang the first wave of wind power development, most of which took place in the United States.

The first modern megawatt sized wind turbine, a 2 MW machine, was tested in Denmark in 1978 and several large machines were also built in the United States during the 1980s, the largest with an generating capacity of 4 MW. In spite of this early innovation, the technology was not yet sufficiently capable and these machines were never deployed commercially. Meanwhile the first recorded wind farm, with 20 turbines, each of 30 kW, was built in New Hampshire, USA in 1980. Further wind farms were built in California, particularly in the rich wind regions of Altamont Pass and Tehachapi, with growth spurred by tax credits. However development slowed during the 1980s as the cost of oil dropped and wind became uneconomical.

The late 1990s and early 2000s saw interest in wind power rise again. This coincided with a growing international effort to limit carbon dioxide emissions from fossil fuel power plants in order to control global warming. Wind was seen as one of the most important alternative sources of energy and new companies entered the industry and began to develop much larger wind turbines. While development was often supported with tax incentives and generous tariffs—and continues to be supported in this way in many countries of the world—costs began to drop, economic performance improved, and the industry gained significant momentum. Estimates suggest that by 2014 there were nearly one quarter of a million commercial wind turbines in operation. In 2015 a wind turbine for use offshore with a generating capacity of 8 MW was launched, the largest yet to be made available commercially.

GLOBAL WIND POWER CAPACITY

The earliest systematic records for global wind power capacity show that after a slow start during the 1970s there were around 10 MW of installed capacity in 1980,[2] most of which was located in the United States. Capacity grew slowly during the 1980s, passing the 1000 MW mark in 1985 but only reaching 1930 MW by 1990. Growth continued into the 1990s with a global capacity of 3000 MW exceeded for the first time in 1994. By 1997, as shown in Table 1.1, the cumulative

[2]Figures are from the Earth Policy Institute.

Table 1.1 Global Annual Wind Power Capacity[3]		
Year	Capacity Added (MW)	Cumulative Installed Capacity (MW)
1997	1530	7600
1998	2520	10,200
1999	3440	13,600
2000	3760	17,400
2001	6500	23,900
2002	7270	31,100
2003	8133	39,431
2004	8207	47,620
2005	11,531	59,091
2006	14,701	73,949
2007	20,286	93,901
2008	26,952	120,715
2009	38,478	159,079
2010	38,989	197,943
2011	40,943	238,435
2012	44,929	283,132
2013	35,692	318,644
2014	51,473	369,597
Source: *Global Wind Energy Council.*		

global capacity had reached 7600 MW. From that date the capacity rose more swiftly with 1530 MW added in 1997, 3760 MW added in 2000, and 11,531 MW added in 2005 when the cumulative capacity reached 59,091 MW according to figures from the Global Wind Energy Council. The rate of growth slowed during the global financial crisis so that the capacity additions in 2010 were virtually the same as in 2009 and in 2013 the total capacity added was less than in 2012. However in 2014 the largest annual capacity addition so far recorded, 51,473 MW, brought the global capacity to 369,597 MW. This compares to a global total generating capacity for all types of power plant of around 5000−6000 GW.

Wind generating capacity is distributed broadly across the globe although some regions have exploited their wind potential more rapidly than others. The figures in Table 1.2 show that the largest regional capacity in 2014 was based in Asia with 141,964 MW installed across

[3]Global Wind Report: Annual Update 2014, Global Wind Energy Council 2015.

Table 1.2 Installed Regional Wind Capacity[4]	
Region	Installed Wind Capacity at End of 2014 (MW)
Africa	2535
Asia	141,964
Europe	134,007
Latin America and the Caribbean	8526
North America	78,124
Pacific region	4441
World total	369,597
Source: *Global Wind Energy Council.*	

the region. This total was dominated by the capacity in China which accounted for 81% of the regional aggregate. Asia was closely followed by Europe with 134,007 MW. Germany, Spain, and the United Kingdom had the largest capacities in this region, with much of the UK capacity offshore. North America had 78,124 MW at the end of 2014; as with Asia, this is dominated by the capacity in one country, in this case the United States which accounted for 84% of the total.

Development elsewhere has generally been slower than in these three regions. Much smaller regional capacities are found in Latin America and the Caribbean (8526 MW) and the Pacific region (4441 MW) although both have good wind regimes. However the smallest regional capacity was in Africa where there was only 2535 MW at the end of 2014. While many of Africa's coastal regions have good wind regimes, the resource is more limited in some inland regions. Even so, the low capacity in Africa is more a reflection of the lack of investment in the continent than of its overall wind power generating potential.

The economic drivers behind the growth of wind generating capacity are the cost of wind compared to alternative sources of electricity, the speed at which a new wind farm can be built—often less than 1 year compared to several years for power plants based on some technologies—and the fact that this is a clean technology. The latter is an important factor driving growth in China where the country is fighting high levels of urban pollution which is mostly associated with coal combustion.

---

[4]Global Wind Report: Annual Update 2014, Global Wind Energy Council 2015.

The global annual growth rate will ultimately be limited by the size of the wind industry and the number of turbines it can manufacture in a year. However annual additions of 50 GW or more are clearly feasible in the current market. With industrial capacity of this scale, the global wind capacity can be expected to climb much higher over the next 25 years.

CHAPTER 2

The Wind Energy Resource

Wind is the movement of air in the earth's atmosphere. These motions that take place within the atmosphere are complex and extremely difficult to predict, hence the difficulty of weather reporting. However, the driving force is relatively simple. Air movement occurs when the air in the atmosphere moves from a region of high pressure to a region of low pressure as it attempts to equalize the pressure, a natural response to a pressure difference. What we refer to as wind is normally the movement of air in the atmosphere close to the ground. This is the wind energy that can be exploited to generate electricity.

The differences in pressure within the atmosphere are generated by energy input into the air and the most important source of this energy is the sun. Solar energy heats the air directly when molecules within it absorb radiation. In addition heat absorbed at the earth's surface helps heat the air close to the surface. This heating effect is not uniform. More heat is absorbed in equatorial regions of the earth which are more directly exposed to the sun's radiation, and less in polar regions which have much less exposure.

When air is heated it expands and rises, creating areas of low pressure, an effect that is particularly pronounced near the equator. The low pressure regions close to the earth's surface in the earth's equatorial regions then pull in air from the colder, higher pressure regions at the poles. Meanwhile the rising hot air flow towards the poles where it mixes with the colder polar air, creating a circulation between the equatorial regions and the poles. This leads to the complex structures of high and low pressure systems that move through the atmosphere, with winds flowing between the high and the low pressure centers.

A more local effect is found where bodies of water such as the earth's seas and lakes absorb heat. Water generally takes longer to heat up than land, but then releases that heat over a longer period so the changes in temperature of bodies of water is generally seasonal.

Wind Power Generation. DOI: http://dx.doi.org/10.1016/B978-0-12-804038-6.00002-5

Land, on the other hand, heats and cools much more quickly and this leads to differential atmospheric pressures between land and sea—sea breezes—over the daily cycle.

There are other complicating factors, one of which is the rotation of the earth. This causes a deflection of the air currents that are flowing from the equator towards the poles as a result of a phenomenon called the Coriolis force. Through all these, the primary energy source driving the wind is the sun so that in effect wind is a form of solar energy.

The complex nature of the interactions that lead to the creation of pressure differences and currents in the atmosphere mean that wind is by its nature both unpredictable and intermittent. Both of these factors are important when considering its exploitation.

Intermittency is a property of the wind. As a result of pressure changes in the atmosphere, sometimes the wind will blow and sometimes it will not. Nothing that humankind can do will change this. The consequence is that sometimes there will be energy to be harvested with wind turbines and sometimes there will not. From a wind power perspective what is important is how often or how much the wind blows. Intermittency is a relatively short time-scale phenomenon. While the wind may be intermittent over an hourly or a daily time-frame, this intermittency will average out so that over the course of each year the wind will generally blow for roughly the same amount of time. Wind measurements taken at a site over a long period, ideally several years, will show how much the wind blows over a 12-month period and from this it is possible to determine the average amount of energy a wind turbine at the site can be expected to produce.

In contrast to intermittency, unpredictability is less a property of the wind and more a property of our perception of it—or of our ability to predict it. Wind is unpredictable to the extent that we cannot predict when it will blow and when it will not. Modeling of the atmosphere is improving year on year and this is allowing increasingly accurate predictions of meteorological conditions, including the wind. Being able to predict the wind accurately on a day-ahead or week ahead time-frame is one of the keys to harnessing wind energy. If it is known when the wind will blow, and when it will not, then alternative energy sources can be made ready for those periods when the wind does not blow. By this means accurate forecasting will allow high levels of wind energy to be supported on a grid system without making the grid unstable.

WIND SPEED AND POWER GENERATION

The key determinant of the wind potential at any given wind site is the average wind speed. Wind speed varies with location. The average will depend on the prevailing wind conditions in the area where the site is located, as well as on local geographical or topographical features. Some regions of the world are simply windier than others and these will yield the best wind energy sites. However, within these windy regions there will be variations in the average wind speed depending on the exact location. The lee of a hill or mountain, sheltered from the prevailing wind direction, will have a lower average speed than the more exposed side of the same hill.

Meteorologists classify wind speed using the Beaufort scale. It was developed for mariners and the levels on the scale all refer to particular wave heights at sea. However, these are correlated with wind speeds. The scale runs from 0 to 12 with zero absolute calm and 12 equivalent to a wind speed of greater than 33 m/s. Winds of this severity are called Hurricane force. Small wind turbines generally operate at between 3 and 7 on the Beaufort scale, equivalent to from 4–5 to 14–17 m/s. Large wind turbines can operate in wind speeds much higher than that, and can be used in winds of level 10 on the Beaufort scale, 25–28 m/s. This represents the limiting wind speed modern turbines can endure. The forces experienced by a wind turbine in winds of greater than around 26 m/s can cause damage to components so wind turbines are generally shut down if the wind exceeds this level. However wind turbines are usually built to be able to survive a wind speed of up to 60 m/s.

Another way of classifying the wind regime, one that was developed for the US wind industry, is based on wind classes. These wind classes were established by the US Department of Energy in order to facilitate the mapping of wind resources. Using this system there are seven wind classes, labeled 1–7. Each is defined in terms of the wind power density at both 10 m above the ground and 50 m above the ground. The ranges of wind power density and wind speed for the seven classes at 50 m above the ground are presented in Table 2.1. The wind power density ranges at 10 m above the ground are half those at 50 m. The wind speed equivalent to each wind power density is also shown in the table.

Table 2.1 Wind Classes, Wind Speed, and Power Density at 50 m		
Wind Class	Wind Power Density (W/m²)	Wind Speed (m/s)
1	0–200	0–5.6
2	200–300	5.6–6.4
3	300–400	6.4–7.0
4	400–500	7.0–7.5
5	500–600	7.5–8.0
6	600–800	8.0–8.8
7	800–2000	8.8–11.9
Source: *US National Renewable Energy Laboratory.*		

Wind developers can use the wind class at a particular site to determine the economic viability of building a wind turbine or wind farm there. Most wind turbines will start rotating when the wind speed exceeds 3 m/s but most developers will want a site with a wind class of 3 or above before they will consider it financially viable.

While the average wind speed or the wind class may be used to develop wind maps, it is important to consider site specific conditions since wind maps are generally related to idealized conditions. In addition, although the average wind speed may be known, this cannot be used to directly calculate the amount of energy that might be available at a particular site because the wind speed varies with time and simply using the average wind speed will result in a prediction of the energy available for power generation that is too high. To obtain an accurate picture the distribution of the wind, the amount of time it blows at each different speed must be taken into account.

The distribution of wind speed over time at a particular site—the frequency at which it blows at a particular speed—has been shown to broadly follow a Rayleigh distribution so this can be used to calculate the actual amount of energy available from the wind using mathematical integration over the distribution of speeds. It is also worth noting that based on this type of distribution the speed at which the wind blows most frequently, called the modal wind speed, is actually lower than the average wind speed.

An additional factor that must be taken into account is that the energy that is available from the wind is not a linear function of the speed but varies as the cube of the wind speed. This means that doubling the wind speed will increase the energy available eight times, so higher wind speeds yield significantly more energy than lower wind speeds. The actual power available in the wind is determined by the equation:

$$P = (1/2)\ CA\rho\ v^3$$

where P is the theoretical power, C is a factor depending on the wind turbine design, A is the area swept out by the rotor, ρ is the air density, and v is the wind velocity. Since theoretical power is proportional to the swept area, increasing the length of the blades will also increase power output; doubling the blade length increases the theoretical power by a factor of four.

While the equation above defines the energy contained within the wind, not all of it can be harvested by a wind turbine. A German scientist, Albert Betz, established in 1919 that the maximum amount of energy that can be extracted from a stream of wind is 16/27 (59.3%) of the kinetic energy contained in the wind. This number is now known as the Betz constant. In practice no wind turbine can convert 59.3% of the kinetic energy from the wind into electricity but some modern machines can reach 80–90% of this value.

Yet another factor when considering wind turbine energy capture is that the wind speed increases with height. The reason for this is that frictional forces close to the earth's surface slow the wind relative to higher regions. This means that a turbine on a taller tower will be exposed to a better wind regime than one on a shorter tower. An empirical law suggests that the wind speed at any height varies relative to the speed at a reference height according to the equation:

Wind speed = Wind speed at reference height $\times$ (height/reference height)$^{1/7}$

This equation takes no account of any ground effects but does offer a reasonably accurate means of generating a wind speed profile from a reference height as height changes.[1] The variation of wind speed with height is also an important consideration in wind turbine design. The rotor of a large wind turbine will experience a greater force on its blades when they are at their highest point than when at their lowest.

---

[1]According to Electropaedia (http://www.mpoweruk.com/index.htm) this relationship was developed at the Pacific Northwest Laboratories in the United States by D. L. Elliott.

This creates a bending force on the turbine shaft which can lead to long term stress damage if the effect is not taken into account.

Wind does not normally move smoothly, particularly close to the earth's surface but instead displays various levels of turbulence. Wind turbulence will reduce the amount of energy that can be drawn from it by a wind turbine as well as generating stress and fatigue in the machinery. Most wind turbulence is caused by obstructions to the wind flow so it is usually greatest close to the ground. Over land the topography can affect the level of turbulence found, as can the type of vegetation at ground level. The marine environment generally generates less turbulence because the sea surface is smoother. Turbulence means that it is better to install a wind turbine on a tall tower so that its blades are kept clear of the most turbulent layer close to the ground. Offshore, turbine blades need to be kept clear of the highest waves.

In spite of all these conflicting effects, wind turbines can generate large amounts of electricity and the stronger the wind, the larger the output. According to the World Energy Council, a 2 MW wind turbine with a hub height of 78 m will generate around 2460 MWh each year at a site with an average wind speed of 5 m/s, 5630 MWh at 6 m/s, and 6725 MWh at 7 m/s. An average wind speed of 5 m/s is typical on inland onshore sites in many regions while 7 m/s is more usually found in coastal regions. Offshore average wind speeds can be higher still. In terms of wind classes, globally the average wind speed onshore is class 1 and that offshore is class 6.[2]

THE GLOBAL WIND RESOURCE

As already noted above, wind is a complex energy source and estimating the size of the global resource is extremely difficult. One estimate, from 1992, put total onshore wind energy at 1,000,000 GW of potential generating capacity.[3] Other surveys have suggested that there is more wind energy to be found offshore than over the world's landmasses. Meanwhile, according to the World Energy Council, all the electrical energy requirements of the European Union could be met

[2]Evaluation of global wind power, Cristina L. Archer and Mark Z. Jacobson, Journal of Geophysical Research, 2005.
[3]World Energy Need, G Cole, Energy World, June, and November 1992.

using offshore wind without going further than 30 km from the shore.[4] All these estimates, and others not included here, indicate that the global resource is far larger than the global demand for electricity.

The wind at ground level—the wind that can be exploited for power generation—is not evenly distributed across the globe. For example, global wind energy maps show that some of the highest winds in the world can be found in southern hemisphere between 40° and 60° of latitude, the region known as the roaring forties. The global wind energy maps from NASA in Figure 2.1 illustrate this. These westerly winds are caused by the movement of air between the equator and the poles as discussed earlier, and by the effect of the earth's rotation. With no landmasses to interrupt them, the winds in the southern hemisphere region resulting from these air movements reach very high speeds. There is a similar region in the northern hemisphere where wind speeds can be high too, but here the major landmasses of America, Europe, and Asia interrupt their movement and wind speeds are not as high as in the south.

Meanwhile speeds at the equator are much lower, on average, which is why mariners have coined the term the Doldrums for the trade winds in these parts. Similarly wind speeds in some equatorial interior regions of the northern part of South America and in the center of the African landmass can be very low.

Global wind maps, where they are available, provide only very rough guidance with regard to the regions with the best wind regimes. More accurate pictures are provided by national wind maps which are available for many countries around the globe. For example, the US National Renewable Energy Laboratory has developed a *Wind Energy Resource Atlas of the United States*. This shows the wind speed at 50 m across the contiguous United States with regions identified by wind class. Large areas of the Midwest as well as parts of the western United States have winds of class 3 or above. Similar maps can be found for Europe and for a variety of individual countries in other parts of the world.

There have also been numerous regional estimates of the wind energy available. For example, the European Environment Agency has

---

[4]World Energy Resources, World Energy Council, 2013.

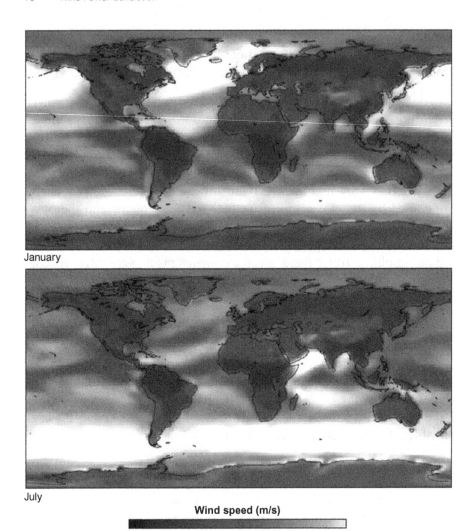

January

July

Wind speed (m/s)

0   7   14

Figure 2.1 Global average wind speed maps for January and July. Source: NASA http://visibleearth.nasa.gov/view.
php?id=56893.

estimated the total wind energy that could be harnessed competitively
in Europe. It found that by 2020 a total of 12,200 TWh would be
available at economically competitive prices, rising to 30,400 TWh by
2030. The latter figure is seven times the expected European demand
for electrical power in 2030.[5] A US study from 2007 found that
national wind potential could support 7800 GW of onshore wind

[5]Europe's onshore and offshore wind energy potential, EEA technical report no. 6/2009, 2009.

capacity and 4400 GW of offshore capacity. China's wind potential has been put at 6960 TWh of energy generated at $75/MWh or lower. The technically exploitable potential in Russia has been estimated to be 6200 TWh.

National wind mapping and wind estimates are usually the starting point when trying to locate good wind energy sites but having identified a potential site, the only way to ascertain whether that particular location will provide an economically viable source of electricity from the wind is to make accurate measurements over an extended period. Sometimes suitable data of this type will already have been gathered by a local agency. In many parts of the world, however, there are neither the resources nor the agencies that can carry out this type of work. Then it falls to a potential wind developer to carry out the necessary wind prospecting.

CHAPTER 3

The Anatomy of a Wind Turbine

Although the modern wind turbine industry was launched during the 1970s, the basic anatomy of the wind turbine was established much earlier. The physical principle upon which all stationary wind machines[1] are based is the use of wind, captured by sails or blades, to turn a shaft to which they are attached, thereby producing a rotary motion that can then be exploited to provide an output useful to humankind. The earliest examples were used either for milling grain or to pump water but from the late nineteenth century onwards wind turbines were also used to generate electricity.

The earliest exploitation of wind led to two basic variants, the vertical axis wind turbine, and the horizontal axis wind turbine and these can both still be found in the modern wind industry. Figure 3.1 shows the two different types schematically. The horizontal axis wind turbine has become the standard configuration for the modern large wind turbine and in this sector there are currently no vertical axis machines. However, vertical variants can be found in the small wind turbine sector.

The horizontal axis wind turbine has a rotor, or propeller, with several blades that capture energy from the wind, causing it to rotate. This rotor is mounted at the end of a shaft that sits horizontally on the top of a tall tower. The shaft is connected to a generator, often through a gearbox, and rotation of the shaft turns the generator, producing electricity. Units following this basic architecture with capacities of several megawatts form the backbone of the utility wind industry.

The vertical axis wind turbine, in contrast, has a vertical shaft with a generator attached to it at ground level. Vanes are mounted onto the vertical shaft and these intercept the wind, causing the shaft to turn. Megawatt-sized vertical axis turbines have been developed in the past

---

[1]As opposed to vessels that use the wind to drive them across the surface of the earth.

Wind Power Generation. DOI: http://dx.doi.org/10.1016/B978-0-12-804038-6.00003-7

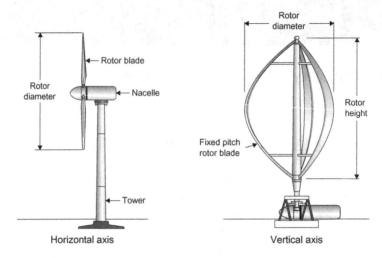

Figure 3.1 Horizontal and vertical axis wind turbine configurations. Source: American Wind Energy Association (taken from Wind Power, The University of Alabama, ME 416/516 The figure was available in 2003 but does not appear to be available from AWEA anymore.

but these have not proved commercially successful. However, vertical axis wind turbines do still find a place in the small wind turbine market. In addition some companies are re-examining the vertical axis concept for offshore applications.

VERTICAL AXIS WIND TURBINES

The vertical axis wind turbine is the earliest recorded wind turbine. Machines of this type, with a vertical axis and cloth sails, were used in Iran and Afghanistan in the ninth century for milling and pumping duties. One of the first recorded wind turbines for power generation, that built by James Blyth in 1887, also had a vertical axis and cloth sails forming a rotor of 10 m in diameter.

The vertical axis concept can be developed in a number of ways but the most distinctive vertical axis wind turbine design is that by the French engineer Georges Darrieus which he patented in 1931. His design has a vertical shaft supporting two thin, curved aerofoil blades, each in the shape of a bow, with the ends of each blade attached to the top and bottom of the shaft, as shown in Figure 3.2. The movement of these blades against the direction of the wind generates an

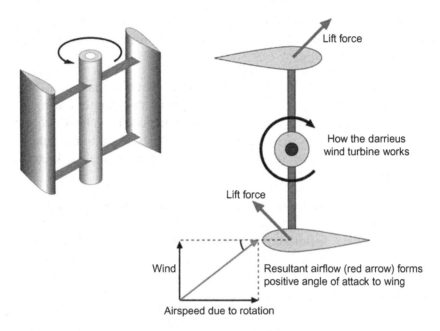

Figure 3.2 Operation of a Darrieus wind turbine. Source: https://en.wikipedia.org/wiki/Darrieus_wind_turbine#/media/File:Darrieus.jpg.

aerodynamic force[2] that acts about the shaft, causing the rotor to rotate. When the same blade has turned through 180° and is moving in the direction of the wind it also generates an aerodynamic force that acts in the same direction about the shaft, again aiding its rotation. A lesser force is generated in positions in between the two extremes of the rotation, into and with the wind. The principles of operation of the Darrieus wind turbine are illustrated in Figure 3.2.

A Darrieus wind turbine will not normally start itself when the wind blows, unlike a more conventional horizontal axis turbine, but once started it will continue to turn while the wind blows, from whatever direction the wind comes. This insensitivity to wind direction is one of its advantages. The speed at which the rotor turns is not related to the wind speed and can be very high, generating large centrifugal forces acting through the blades onto the shaft. The curved, "eggbeater" shape of the Darrieus turbine blades helps make them

---

[2]The Darrieus wind turbine exploits the aerodynamics of linear motion whereas the horizontal axis wind turbine exploits the aerodynamics of rotary motion of a propeller. The latter cuts a flat disc shape through the air passes when crossing the plane of the blades. The vertical axis turbine creates a shape more akin to a cylinder through which air passes.

self-supporting when under centrifugal stress, minimizing the need for additional support structures to prevent them distorting out of shape when rotating, or failing.[3]

One problem with this type of turbine is that the maximum force generated by the blades in the wind occurs as two points during the rotation of each blade, generating a pulsing power cycle which can lead to a resonant frequency at particular rotational speeds. Resonances of this type are capable of causing the blades to break in extreme cases.

Against this, the positioning of the generator at ground level means that most of the weight is low, so the center of gravity of the machine is low. A horizontal axis machine has most of its weight at the top of the tower. In addition, whereas with a horizontal axis wind turbine all the force of the wind on the blades is transmitted to the top of the tower, so all this force acts as a bending moment on the tower top, with the Darrieus wind turbine the force is borne equally at the top and bottom of the shaft. For both these reasons, the Darrieus wind turbine is potentially cheaper to construct than a conventional horizontal axis machine Figure 3.3.

There are some other major disadvantages, however, which have so far outweighed these advantages. One is that much of each blade is too close the ground or sea surface to exploit the best wind conditions. The blades are more expensive to construct than conventional blades and finding blade materials that can withstand the pulsing power cycle has proved difficult. Many of these problems came to light when the Darrieus design was revived during the 1970s, particularly by the National Research Council of Canada and by engineers at the Sandia National Laboratories in the United States. This revival led to the construction of the largest Darrieus wind turbine ever built, called Éole. Éole is a 4 MW vertical axis wind turbine with a rotor diameter of 64 m and a height of 96 m. It was built on the southern shore of Lake Lawrence in Canada where it started operating in 1987 and continued until 1993 when its main bearing was damaged. The turbine still stands, a monument to the renewable energy drive of that period and has since become a tourist attraction.

---

[3]The form of the blades is that of the troposkein, the shape that a rope forms when anchored at its ends and spun. The shape leads to the minimum stress.

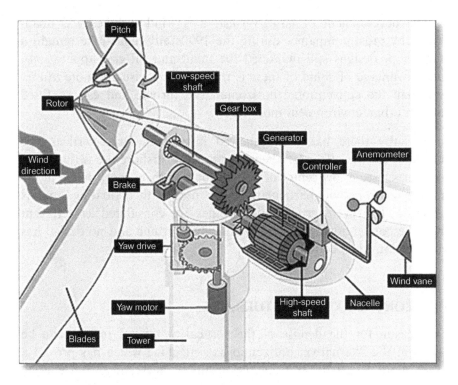

Figure 3.3 The principle components of the drive train and nacelle of a horizontal axis wind turbine. Source: US National Renewable Energy Lab (taken from http://windeis.anl.gov/guide/basics/).

The Darrieus is one of a number of vertical axis wind turbine configurations. Another, also patented by Darrieus himself is the H-shaped rotor in which the blades of the rotor are vertical but straight so that the whole rotor has the shape of a capital H. This type of rotor is called a Giromill. It is simpler to build than the eggbeater design but requires much stronger supports for the blades to withstand the centrifugal forces.

A further development of this type of vertical axis turbine is the cyclo-turbine. This also has vertical blades like the Giromill but in this case each blade can be moved about a vertical axis through its nose so that the angle of attack can be changed as the rotor turns. This allows the aerodynamic forces to be optimized at different positions about the rotational axis, reducing or removing the pulsing nature of power capture when the blades are fixed. In addition this type of vertical rotor is self-starting. However the price for this is a much more complex rotor mechanism. A system is required to sense the wind direction too, as the optimum blade angle of attack at any position will vary with wind direction.

The development of large vertical axis wind turbines was abandoned by most companies during the 1990s although there remain a number of designs still marketed for small and off-grid applications. One advantage of some of these is that they are visually more attractive than the conventional horizontal axis turbine and can be fitted into an urban environment more easily.

Recently there has been renewed interest in large vertical axis machines for use offshore. The advantage for offshore use is that these machines have a low center of gravity making them more stable in the marine environment, particularly when used with a floating support structure. However vertical axis turbines are considered less efficient than the conventional horizontal axis wind turbine and no design has yet been launched commercially.

HORIZONTAL AXIS WIND TURBINES

The reason for the demise of the vertical axis wind turbine can be found in the alternative, horizontal axis machine which has proved a much more successful and reliable means of harvesting wind energy. The workhorse of the wind industry today is a three-bladed, upwind, horizontal axis wind turbine mounted onto a tubular tower. This design has emerged over the past 20 years as the standard for large single turbine and wind farm installations both onshore and offshore. Before that, in the period between the late 1970s and the beginning of the twenty-first century, a large number of variants were tested. Today, while the basic format is standard, there remain variations in the drive trains and blade construction as well as new approaches to tower construction that are being tested to make onshore installation of very large turbines easier.

The standard utility wind turbine has a rotor with three blades because this is considered the optimum compromise between balance and cost. Other rotor types have been built in the past including turbines with one, two, three, four, and, in rare cases, even more blades. The more blades there are the more evenly the rotor is balanced. In addition, the rotor speed for ideal energy capture will be lower, the more blades it has and this can, in principle at least, allow larger turbines to be built because centrifugal stress is lower at lower rotational speed. Within certain limits, more blades can capture more energy.

The downside of rotors with many blades is the cost. Blades are complex and expensive items so the fewer there are, the cheaper the wind turbine. The extreme, a rotor with one blade, has been tested but has not proved practical; two blades can achieve better rotor balance but still create problems, hence the standardization on three blades.

The speed at which the rotor turns is determined primarily by the wind but there must be a means of controlling this. Controlling the rotational speed is accomplished either passively, using a blade with an aerofoil shape that will automatically slow if the wind speed becomes too high, or by using moveable elements in the blades that allow the amount of energy taken from the wind to vary as speed varies. A turbine rotor also needs a braking system that can be used to stop it rotating completely, particularly in very high wind conditions.

The rotor is mounted on one end of a drive shaft, the other end of which is connected to a generator. There may be a gearbox between these two elements to match the rotational speed of the rotor, which is usually very slow by generator standards, to the rotational speed required by the generator to produce an alternating current and voltage at grid frequency. The principle components of the horizontal axis wind turbine drive train are shown in Figure 3.3. There can be wide variations in this drive train as designers attempt to manufacture more efficient and more reliable wind machines. These may include omitting the gearbox altogether so that the rotor drives the generator directly and the use of some form of variable speed generator that can extract the optimum amount of energy from the wind no matter what the wind speed. Early wind turbines used modified electric motors as their generators. These could not maintain grid frequency on their own and required the grid to control the speed at which they turned. However, modern turbines are required to be able to maintain grid frequency independently of the grid.

The drive train is housed in a structure called the nacelle which sits at the top of the wind turbine tower. The nacelle protects the components from the weather. The complete tower-top structure is attached to the tower through a bearing that allows the nacelle to rotate about a vertical axis so that the rotor is always facing into the wind. In order to maintain its orientation as the wind direction varies, the nacelle must be equipped with a yawing motor that moves the tower-top structure as necessary. Many earlier wind turbines used a rotor that faced downwind. This allowed the wind to control the orientation so no yaw-ing motor was necessary. However, this type of design can increase both noise from the turbine and fatigue stress which is why the upwind orientation is preferred.

The tower upon which the nacelle is mounted is normally of tubular steel construction today. The tower will taper so that the base has a larger diameter than the top. Other construction methods are possible, including the use of concrete although this has usually proved more expensive than steel. During the early years of wind turbine development lattice steel towers were also common but these are rarely found today except for small capacity machines. Small wind turbines may also use other tower structures such as tripods, or poles and guy-wires. The tower of a large utility wind turbine will contain a lift to reach the nacelle and there will be a transformer at its base to raise the voltage of the power from the machine before it is fed to the grid, usually via a local substation.

The turbine and tower require a stable foundation to prevent the force of the wind toppling the structure. On land it is normally a mas-sive concrete structure, often with steel reinforcement, that relies on the strength of the soil in which it is buried to withstand the forces exerted by the wind on the tower. More complex foundations are often used for offshore structures, including monopiles that are driven into the sea bed and multilegged supports such as tripods.

Although single wind turbine installations are common the largest number of large wind turbines are to be found in wind farms. These are arrays of wind turbines arranged so that they can capture wind energy efficiently over a large area of land without interfering with one another. The dynamics of wind flow through and around turbines needs to be analyzed to optimize the layout of a wind farm. A large wind farm will act as a single power station, with the power from each turbine collected at a local substation before being fed into the grid. Since areas with good wind regimes are often far from the centers of electricity demand, and therefore not located close to the backbone of a grid system, large wind installations will often need a special connec-tion to the grid that can support its output. In regions where there are large numbers of wind farms because of the particularly good wind resource available, utilities are beginning to establish special wind grids to transport this energy.

Table 3.1 Average Wind Turbine Size	
Year	Average wind turbine size (kW)
1984	30
1989	150
1993	226
1996	642
1998	750
2000	800
2002	1100
2005	1300
2009	1600
2012	1800
Source: *Composites World, Riso and Force Energy, Wind Systems, Navigant.*	

Wind turbine size has grown steadily since the 1970s. Table 3.1 shows how the average size of wind turbines being installed grew between 1984, when it was around 30 kW, and 2009, when the average size of units installed was 1.6 MW. In 2012 the average turbine size was 1.8 MW. Since then the average size has grown even larger so that in the middle of the second decade of the twenty-first century it is probably between 2 and 3 MW. The average size of offshore wind turbines was 4 MW in 2014. Average size will grow higher still, particularly offshore, as machines with individual generating capacities approaching 10 MW are unveiled. Economies of scale mean that larger machines are generally cheaper so this trend is likely to continue unless some natural limit is reached or size becomes a factor.

CHAPTER 4

Rotors and Blades

The rotor of the wind turbine is the element of the machine that captures energy from the wind and its design is crucial to the efficiency of electricity generation. The wind turbine designer has a number of parameters and variables to play with when trying to achieve the optimum design. The number of blades that make up the rotor, the shape of the blades, their length, and the speed of rotation of the rotor all affect the amount of energy that can be taken from the wind.

A blade captures energy from the wind as it sweeps across the flow of the wind. For a horizontal axis wind turbine the rotor blades sweep out a circle—often called the swept area—perpendicular to the direction in which the wind is blowing. In order that each segment of the circle swept out by the rotor is exploited as the wind passes through it, a blade must interact with it. If the blades turn too slowly, some air will pass untouched. Too fast and the turbulence from one blade is likely affect the next, which is counterproductive. More blades can make it easier to exploit all the energy carried by the wind but too many, and the blades start to interfere with one another. In addition, the more blades there, the thinner they must be for optimum efficiency and this can affect their structural integrity.

Once the blades of the rotor have captured energy and transferred it to the wind turbine shaft as rotational motion, this rotation must then be converted into electrical energy. The rotation-to-electricity conversion is carried out using a generator attached to the end of the shaft. In order to maximize energy capture, generators for wind turbines have become increasingly specialized. Conventional generators in power plants connected to the grid operate at constant speed, synchronized with the grid. However the demands of wind generation means that some wind turbine generators can operate at two different speeds while others are designed to operate at any speed.

Ideally the complete drive train, comprising rotor, generator, and any intervening gearbox, operates as a single unit. Getting the design of this correct is the key to both efficiency and reliability of wind turbines.

Wind Power Generation. DOI: http://dx.doi.org/10.1016/B978-0-12-804038-6.00004-9

BLADE LENGTH

The length of a wind turbine blade depends on the size of the turbine, on the specific blade design and on the site where it is to generate power from the wind. A 3 MW wind turbine might have a blade length of 45 m at one site where the wind regime is very good and a blade length of 55 m in order to produce the same amount of energy at a different site. The size of commercially available wind turbines has increased over time, as illustrated in Figure 4.1.

The amount of energy captured by the blades depends on the swept area and this increases as the square of the blade length so a small increase in blade length can produce a large increase in energy capture. Wind turbines in the 2–3 MW generating capacity range will typically have blades that are between 45 and 55 m long. A commercial 7.6 MW wind turbine has blades that are close to 65 m long. Meanwhile there is a company that is developing blades of up to 100 m in length. This would probably be adequate for a 10 MW wind turbine. A design to test the feasibility of a 20 MW wind turbine suggested a blade length of 125 m would be needed, giving a rotor of 250 m in diameter.

ROTOR AERODYNAMICS

In early windmills the sails acted like the sails of a ship, catching the wind which then pushed them around in the same way as a ship is pushed along. While effective, this is not the most efficient way of

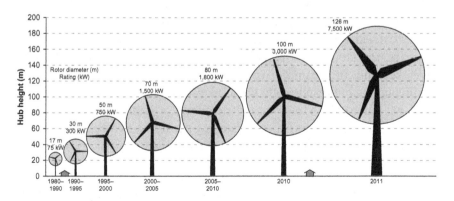

Figure 4.1 The evolution of wind turbine rotor diameter and hub height. Source: https://en.wikipedia.org/wiki/History_of_wind_power#/media/File:Wind_turbine_size_increase_1980-2011.png.

using wind energy. Better performance can be achieved by shaping the blades like the wings of an aeroplane.

The blades of a horizontal axis wind turbine have a cross-section that is the shape of an aerofoil, the same shape that is used in an aeroplane wing to create lift. The aerofoil shape is such that air flowing over the top surface of a wing has to travel further than air traveling over the lower surface. The practical result of this is to create a lower air pressure over the top of the wing than under it. This differential pressure creates a force pressing upwards on the wing, a force which is generally called lift, as shown in Figure 4.2.

The lift generated by an aeroplane wing is used to raise it into the air. The same force operates on a wind turbine blade to transfer energy to the rotor.

There is a second force that is generated on a wing or aerofoil in a moving current of air, a force called drag. This is essentially the resistance of the aerofoil to the airflow, the force that any body experiences when exposed to the wind, and its effect is to try and slow the aerofoil. The drag acts perpendicular to the lift on an aerofoil.

When an aerofoil is pointing directly into the wind[1] it presents the smallest cross-section into the wind and so the drag is at its lowest. However this does not create the maximum lift. Raising the front of the aerofoil relative to the wind direction, called increasing the angle of attack, increases the amount of lift that the aerofoil creates but it

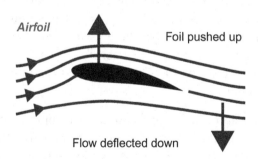

Figure 4.2 The aerofoil principle. Source: Wikipedia commons (https://en.wikipedia.org/wiki/Lift_%28force% 29#/media/File:NASANewtons3rdGlennResearchCenter.gif).

[1]The direction of the aerofoil is usually defined by drawing a line from its tip to the point of its nose.

also increases the drag. Eventually, if the angle of attack is increased too much the air flowing over the top of the aerofoil creates such a low pressure region that the airflow across the top becomes turbulent. This causes the amount of lift to drop of rapidly, an effect in an aeroplane known as stalling. All these effects are of importance when considering the blade of a wind turbine.

If a wind turbine blade is thought of as an aeroplane wing, then when the blade is stationary the lift that it experiences as the wind passes over it acts perpendicular to the blade, in the plane of the circle swept out by the blades, and generates a force about the shaft of the turbine rotor—a torque—causing it to turn. All the blades have the same shape so that each one contributes a force in the same rotational direction, no matter what position it is in. Under these circumstances all the drag on each blade would be felt as a force parallel to the shaft, pushing the rotor assembly against the tower.

When the rotor is turning, the situation becomes more complicated. The movement of the blade across the wind direction means that the blade experiences the wind as coming from a different direction to the actual wind direction, a direction called the apparent wind direction which can be found by vector addition of the wind speed and the speed at which the blade is moving. The effect of this is to shift the actual direction in which both lift and drag are felt so that some of the lift appears as a force parallel to the wind turbine shaft, pushing against the tower, while part of the drag reduces the actual torque felt by the shaft.

In consequence of this the best aerodynamic performance is found when the aerofoil blade is turned towards the apparent wind angle. In addition, the actual speed of the blade as it turns increases from root to tip. To account for this the blade aerofoil must be twisted along its length. Typically a twist of $10°-20°$ is required to obtain the best aerodynamic performance.

The optimum aerofoil shape for a wind turbine blade is relatively thin. While this would extract the most energy from the wind, the blade also has to be strong enough to withstand the forces of lift and drag, the gravitational force acting on it as a result of its weight and the centrifugal forces when it is rotating. This means that the blade must be thicker than the optimum and the thickness must be greatest near the root where the greatest strength is required.

ENERGY CAPTURE

A wind turbine is an obstacle in the path of the wind and like any obstacle it has an effect on the wind flow. The degree to which it actually presents a physical obstacle is defined by a term called the solidity of the turbine which is the total area of the blades in the direction of the wind as a proportion of the area swept out by the blades. There is an optimum solidity, which varies with the tip speed of the blades[2] and for practical wind turbine rotors it is only a few percentage points. This limits total blade thickness and the more blades there are, the thinner they need to be. This is one reason why the optimum is considered to be three blades.

The wind approaching a wind turbine rotor will start to slow even before it reaches the rotor because of the obstacle that it presents and as the air slows, some of it will flow around the turbine rotor instead of through it. The Betz limit, discussed in Chapter 2, means that only a proportion of the energy in the wind that finally flows through the rotor can actually be captured by a wind turbine. The Betz limit suggests that at most 59% of the wind energy can be captured. In terms of wind speed, this translates in the ideal situation into a speed of the wind that has just passed through the rotor of around one third of the speed of the wind before it interacted with the turbine. Practical wind turbines can extract up to 50% of the energy in the wind.

There are ways of capturing more energy from the wind. One of these is to encase the turbine in a shroud that has a diameter greater than the wind turbine itself. The effect of this is to funnel more air through the turbine. This increases the energy capture by the turbine. However if the diameter of the device is now considered to be the diameter of the shroud rather than the simple turbine diameter, then the overall efficiency is actually lower than it would be for an unshrouded turbine of this larger diameter. It then comes down to a matter of cost. Is it cheaper to build a small turbine with a large shroud, or to build a larger turbine without a shroud?

There is one further factor that has to be considered when designing a turbine blade for optimum energy capture. The wind that leaves the disc created by the rotor should be traveling at the same speed whether

---

[2]The tip speed is the tangential speed at which the tip of a blade passes through the air.

it passes by the center of the rotor, near the blade root, or passes close to the tip of a blade. If not, turbulence will be created downwind of the rotor and this will lead to a loss in efficiency. In order to make sure that the wind velocity remains uniform across the swept area, the blades must narrow towards their tips. This reduces the lift generated from each section of blade, the further it is from the root. However, since the further it is from the root, the faster it is moving through the air, the overall energy extraction at each section is the same and so the reduction in wind speed as the wind crosses the turbine rotor is the same.

TIP SPEED RATIO

The speed at which a wind turbine rotor turns is determined by the tip speed ratio, the ratio of the speed at which the tips of the blades move through the air to the wind speed before the wind has been slowed by the turbine. A high tip speed ratio—implying that the rotor is turning quickly—will shift the aerodynamic force acting on the turbine blade further away from the ideal at which it acts as a torque about the axis so in principle a low tip speed ratio would appear ideal.

In practice the tip speed ratio cannot be too low because otherwise it reduces the overall efficiency of the turbine. This can be understood in terms of how the blades interact with the wind. If the rotor turns very slowly, some of the air will pass between the blades without interacting with them. The energy contained in this air will then be lost.[3] At the other extreme, if the blades move too fast, then one blade passing through the air will create turbulence which will be experienced by the following blade arriving too soon in the same air space. This leads to a loss of efficiency too. So the optimum is that each blade should pass through the air and the turbulence it creates should have dissipated before the next blade arrives in the same position.

---

[3]There is an alternative way of looking at this. The power generated from a wind turbine blade is equal to the product of the force generated by lift and the speed at which the blade is moving. If the blade moves more slowly, then in order to generate the same amount of power it needs to be wider so that it can generate more lift. Wider blades will interact with more air than thin blades and so extract the power that would slip between thin blades. However, wider blades are both more expensive to construct and also lead to greater tip losses. As a consequence higher speed and thinner blades are considered preferable. This leads to a higher tip speed ratio.

There are additional tip losses where air from the upwind, high pressure side of the turbine blade slips round the tip and so does not contribute to energy capture. This is greater, the smaller the tip speed ratio. Both of these effects mean that a higher tip speed ratio offers the optimum in terms of cost and efficiency.

As with many aspects of wind turbine design, the actual tip speed ratio chosen is usually a compromise. Typically it will be between 7 and 10 for a three-bladed design which is slightly higher than the optimum from an energy capture perspective. If the turbine is designed to reach its optimum power output in a wind speed of 12−15 m/s, typical of many utility turbines, this implies that the blade tips will be moving at around 120 m/s. Since tip speed will depend on the length of the blade, this suggests that the longer the blade of a wind turbine, the lower the overall rotational speed for a given tip speed ratio.

These are also factors that will limit the absolute tip speed. They include centrifugal forces resulting from rotation, which act at the root of the blade, and the amount of erosion of the blade by water in the air when it moves at very high speed. Bird impacts can increase at high tip speeds and the noise made by a wind turbine rotor depends critically on the tip speed too. Too high, and the noise becomes intrusive.

SPEED CONTROL

A wind turbine will be designed to operate under a range of wind speeds. It will normally start to turn at around 3 m/s and then the power output will increase as the wind speed increases and the rotor turns more quickly until the wind speed reaches somewhere between 12 and 15 m/s when it reaches its design power output. If the wind exceeds this wind speed then there has to be a way of enabling the rotor to reduce the total amount of energy it captures in order to keep output at the rated power. Finally, if the wind speed exceeds perhaps 25 m/s, the cut-out speed, then the turbine will be in danger of becoming damaged and must be shut down.

One way of controlling the speed is to design the aerofoil shape of the blades so that once the wind speed exceeds the design maximum the blades begin to stall. They will then shed lift gradually as the wind speed increases and keep the total power extracted from the wind roughly constant. This type of passive speed control does not provide the perfect speed management as the rotor speed will still increase slightly as the wind speed increases above the design speed. In order to cater for this the turbine will usually operate at slightly below its maximum.

The main alternative to passive control is an active system. This involves building blades that can be rotated about an axis along the length of each blade. This allows the angle of attack of each blade to be varied with wind speed. Using this method, once the rotor reaches the speed at which maximum power is generated then the blades are "feathered," reducing the angle of attack as wind speed increases to reduce the lift generated to keep the power constant. This same mechanism can be used to stop the turbine completely if wind speed becomes too high.

Instead of reducing the angle of attack when the wind speed exceeds the rated speed for the design, it is also possible to increase the angle of attack at this point until the blade starts to stall. This will have the same effect, though the forces generated on the blade will be different. This type of speed control is also less sensitive to gusting because when the blade is stalling, a gust will make it stall more. With the alternative, reducing the angle of attack, a gust will increase the rotor speed before the active system can compensate.

BLADE DESIGN AND CONSTRUCTION

A wind turbine blade must extract as much energy as possible from the wind but at the same time it must be strong enough to withstand all the forces that act upon it as it turns in the wind. In order to achieve this a turbine designer will develop an ideal aerofoil shape for the whole blade length. Then this optimum aerofoil form will need to be modified to accommodate the demands of maintaining structural integrity. The rotor and blades will normally be designed to provide best performance at the rated wind speed but the blades will also have to work successfully at other wind speeds. Given all these demands, the final design of any wind turbine blade is bound to involve a number of compromises.

Once the compromises have been made and the design finalized, the overall shape can be generated with great accuracy using computer modeling and computational fluid dynamics. However, that presents a further problem because the resulting blade shape is likely to be complex and therefore expensive to build unless further simplifications—further compromises—are made. Practical blade construction is therefore something of an art as well as a science.

There are a number of forces that act on a wind turbine blade and on the whole rotor. The aerodynamic lift generates a force that is ideally perpendicular to the blade in the plane of the swept area. This force acts along the whole length of the blade, creating a bending force that is greatest at the root of the blade. The force will bend the blades in the place of the swept area. Meanwhile aerodynamic drag will generate a bending force that acts perpendicular to the plane of the swept area and parallel to the shaft. The weight of the blade also generates a force that acts along the blade and again is greatest at the root. Then there is the centrifugal force generated when the rotor rotates. This again will be most strongly felt at the root particularly where the blade is attached to the shaft.

Other forces are more unpredictable. These include the forces generated by gusts of wind and by turbulence which can act in any direction. Turbine blades are flexible and they will bend under the forces to which they are exposed. However, the blades must be stiff enough that they can never bend sufficiently to hit the tower as they pass it during each cycle.

To accommodate all these requirements the blades of modern wind turbines are usually made from composite materials that can be designed to confer strength and stiffness where it is needed while avoiding excess weight where it is not. Wood, which has relatively high strength and has a good resistance to fatigue failure has been used in the past for smaller wind turbine blades but it not usually found in the large blades of modern utility wind turbines.

A typical modern blade will normally be hollow, with a load-bearing spar down its length to provide strength and rigidity. On to this is built a much lighter shell which provides the aerofoil shape as well as providing additional strengthening elements, particularly related to twisting of the blade. Figure 4.3 shows a cross-section of a turbine blade, with two glass-fiber reinforced shells strengthened with two cross bracing webs and additional balsa wood stiffening.

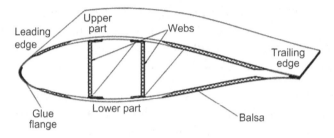

Figure 4.3 Cross-section of a wind turbine blade. Source: L M Windpower http://www.lmwindpower.com/
Rotor-Blades/Technology/Design/Blade-Concept. Courtesy of L M Wind Power.

In order to fabricate the complex shape of a modern turbine blade, most are made from fiber reinforced plastics. These composite materials have a better strength to weight ratio then either wood or metal. In addition the fibers within the composite can be aligned to provide strength in just the directions in which it is needed. So, for example, aligning the fibers in some components along the length of the blade to strengthen it to resist bending forces.

The two most common types of composite used for turbine blades use either glass fibers or carbon fibers. Carbon fibers are around two times as strong as glass and exhibit three times the stiffness. Against that, carbon fibers are significantly more expensive. As a consequence they are only used in the very largest turbine blades, and then only in critical components of the blade where their superior properties are most needed.

ADVANCED BLADE AND ROTOR DESIGN

The standard three-blade upwind rotor used in utility wind turbines will vary from manufacturer to manufacturer but the basic design parameters and constraints, as outlined above, are constant. At the same time some more advanced designs are being developed that may be able to provide better efficiency, lower cost, greater reliability, or all of these at the same time.

While many turbines have active blade pitch control to regulate the rotor speed, this is usually applied equally to each blade. However one advanced idea is to control the blade pitch for each blade independently. This can be particularly helpful to adjust for the change in wind speed with height. For a large wind turbine with a rotor diameter of perhaps 150 m, the wind speed experienced by a blade at the bottom of the swept area will be lower than that at the top. This creates a differential lift and drag from top to bottom of the rotor and this applies a bending force to the shaft of the turbine which can create stress in drive train components, leading to early failure. If blade pitch is adjusted during each cycle, the aerodynamic forces can be balanced across the swept area, reducing this stress.

Another advanced blade design technique involves designing blades that can twist under very heavy loads such as experienced in gusts. This allows them to shed load, reduce lift and drag, and therefore to reduce the level of fatigue stress along the blade. A more complex approach to the same problem is to build blades with a number of adjustable elements, controlled by microprocessors, so that the overall blade shape can be modified in much more sophisticated ways to cater for different wind conditions.

Yet another advanced blade design involves building blades that include a telescopic section so that the blade length can be changed to suit wind conditions. A large rotor can harvest more energy when the wind is light but the same large rotor will also experience high fatigue loads in high winds. If the rotor diameter can be reduced in high winds and increased in low winds, then rotor speed can easily be controlled while the fatigue loads are significantly reduced. All these advanced rotor and blade designs involve more complex structures so cost becomes a determining factor. If an advance can make a wind turbine more reliable or improve its efficiency, then this may be cost effective over the lifetime of the turbine.

TURBINE LIFETIMES

Modern utility wind turbines are designed for a life of 20–25 years similar to that of many conventional power plants. Depending on the rate of advance of turbine technology, and particularly when turbine sizes are increasing significantly with time, it may be cost effective to replace a small turbine with a much larger one long before it has reached its design age. Repowering of this type has taken place during the first decade of the twenty-first century. However as sizes stabilize, it becomes more likely that turbines will operate for their full lives.

As with most machines, wind turbines age, they become less reliable and maintenance costs rise with time. In addition, output drops.

A study of 282 wind farms in the United Kingdom found that they lost 1.6% of their output each year and the average load factor dropped from 28.5% in the first year to 21% in the 19th year. This was equivalent to a loss of 12% of the turbine's output over a 20 year lifetime, and an increase in the levelized cost of electricity of around 9%.[4]

[4]How does wind farm performance decline with age? Iain Staffell and Richard Green, Renewable Energy, Volume 66, pp 775–786, 2014.

CHAPTER 5

Drive Trains, Gearboxes, and Generators

The drive train of a conventional horizontal axis wind turbine has the hub of the rotor at one end, connected to the main shaft. At the other end of the shaft is a generator that converts the mechanical rotary energy in electrical energy in the form of an alternating current. There may be a gearbox between the rotor hub and the generator to match the rotational speed of the rotor to that of the generator.

The rotational speed of a large utility wind turbine is likely to be 20 rpm or less. A conventional two-pole generator synchronized to a 50 Hz grid must turn at 3000 rpm, or 3600 rpm for a 60 Hz grid. For a four-pole generator the speed is 1500 rpm or 1800 rpm respectively. In the conventional layout, therefore, the gearbox must increase the speed by between 75 and 150 times.

The gearbox has often been considered the weak link in the drive train, the component most prone to failure. The shaft of a wind turbine is subject to bending forces and these can have a profound effect on a gearbox. Premature failure has been common. Gearboxes can be strengthened but an alternative is to abandon the gearbox and connect the hub directly to the generator. This direct drive system simplifies the drive train but then requires a special, higher cost, generator that can generate grid frequency power from the very slow rotational speed of the rotor. These generators are much heavier than conventional generators.

Another consideration with a wind turbine is that the speed of the rotor will vary constantly as wind speed changes and yet a conventional generator must turn at a fixed speed if it is to synchronize with the grid. There are systems, as discussed in Chapter 4, to control the speed but this may be at the expense of maximum efficiency, with some energy being shed in order to maintain the correct speed. An alternative solution is to use a variable speed generator instead of a fixed speed generator. Again this will involve higher cost but may be cost effective if the efficiency and reliability gains are large enough.

Wind Power Generation. DOI: http://dx.doi.org/10.1016/B978-0-12-804038-6.00005-0

THE WIND TURBINE GEARBOX

The gearbox in a wind turbine drive train must increase the rotational speed of the rotor to match that required by the generator. Conventional generators for power generation are normally two-pole and four-pole machines that rotate at 3000 rpm or 1500 rpm, respectively, when synchronized to a 50 Hz grid (3600 rpm and 1800 rpm for 60 Hz).

Very small wind turbines can rotate at speeds compatible with these conventional types of generator and so can be connected directly to the generator. However, larger, megawatt class machines rotate much too slowly to be able to match the speed of any form of conventional generator. The traditional solution has therefore been to place a gearbox in the drive train that will step up the rotational speed, by around 100 times for a typical megawatt class wind turbine. The precise ratio will depend on the turbine design.

For utility wind turbines up to around 1.5 MW in capacity, inserting a gearbox into the drive train offers the most economical solution to the problem of matching rotor and generator speeds. However, for larger wind turbines, over 3 MW in capacity, there is an alternative, the use of a direct drive generator with no gearbox in the drive train. These generators are more expensive but are becoming common in large utility turbines. However there are still arguments in favor of using a gearbox, and ways of improving gearbox reliability are being explored by some manufacturers.

It is possible to build generators that can operate at much lower speeds than a conventional generator, though still faster than the speed of the rotor. These have more poles and require a higher torque to turn them, so they are also more expensive. However, the gearbox needed to raise the speed to perhaps 750 rpm or 375 rpm will be simpler, with fewer elements, and will potentially be more reliable. Most of the gearboxes in large wind turbines up to around 1.5 MW in capacity are based on planetary or epicyclic gearing systems. These are often extremely complex. However the combination of a low speed generator and simpler gearbox with a lower step-up ratio may be able to compete with the direct drive generator both for reliability and cost. A three stage gearbox for a medium speed drive train is shown in Figure 5.1.

Figure 5.1 Three stage wind turbine gearbox for medium speed drive train. Source: Generator Gearbox GPC840D, differential design, Bosch Rexroth AG. Rexroth Bosch Group. The image was taken from this brochure http://dc-corp.resource.bosch.com/media/general_use/industries_2/renewable_energies_6/windenergy/BRW_Wind_Lay09_AE_FINAL.pdf.

The main difficulty with gearbox design is that wind turbine gearboxes have to withstand a range of shocks and loads that would not be found in most gearbox applications. As well as the bending forces on the shaft which are transmitted into the gearbox there are sudden changes in torque caused by gusting and even reverse torques under severe conditions. Therefore, the design of the drive train must either insulate the gearbox from these shocks or the gearbox must be built to withstand them.

One partial solution to reduce load and wear is to separate the means of support of the turbine rotor and shaft from the transmission of the turbine torque. This helps isolate the gearbox from some of the forces experienced by the rotor. Meanwhile one of the main causes of gearbox failure is misalignment of the shaft as a result of the forces to which it is exposed. Using a stronger chassis structure for the drive train can reduce the size of the problem. Regular realignment of the drive train will help minimize the stress and wear within the gearbox.

Different manufacturers choose different solutions for wind turbine drive trains but many still prefer to use some form of gearbox. As the industry has matured so the main problem areas within gearboxes have been identified but it is still difficult to eliminate these problems. Instead gearboxes are being designed to withstand the particular stresses, shocks, and wear to which they are exposed. In addition, a continuous monitoring and maintenance regime can help extend gearbox life by identifying problems before they have become serious.

Since many of the problems identified with gearbox wear and failure are related to the size of the forces to which they are exposed, another solution that has been examined is to reduce the torque by splitting the drive train into a number of smaller units, each with its own gearbox and generator. Rotors driving 8 or 16 separate gearboxes and generators have been tested, with varying success.

Even with the advances in gearbox design, the wind turbine gearbox continues to be a high maintenance unit. While a wind turbine will be built to operate for 20 years, the gearbox is likely to require a major overhaul every 5 years.

GENERATORS FOR WIND TURBINES

The generator in a wind turbine is the electromechanical machine that converts mechanical energy into electrical energy. As such it is one of two vital energy conversion components, alongside the wind turbine rotor, that between them convert energy in the wind into electricity.

The first commercial wind turbines of the 1970s and early 1980s often used very simple asynchronous, or induction generators to produce electricity. This type of generator has a conventional stator with three coils that provide the three phases of a grid electricity supply. However the rotor is either a wound rotor, or an alternative robust design called the squirrel cage. In both cases the rotor coil is a closed circuit.

When an induction generator is connected to the grid the alternating current in the stator windings generates a magnetic field which induces a large current in the rotor. This produces a magnetic field which acts against the magnetic field produced by the stator and so the rotor begins to rotate. As the rotor speed increases so the current induced in the rotor falls until an equilibrium point is reached with a small induced torque when the speed of rotation is just below that of the grid frequency. The difference in speed between the alternating magnetic field in the stator windings (at grid frequency) and the rotational speed of the stator is called the slip. As a result of this slip, the induction machine is never synchronized with the grid, hence its other name, the asynchronous generator.

When an external torque from a wind turbine rotor is applied to an induction generator that is already connected to the grid, and rotating, the additional rotational energy raises the speed of the rotor above that of the rotating magnetic field in the stator generated by the grid frequency. This creates a negative slip and the induction machine starts to generate a current within the stator coils which is fed back into the grid. The peak power output is usually achieved within $1-5\%$ over the grid frequency.

One advantage of the induction generator, apart from its simplicity, is that the speed at which it rotates will increase or decrease as the torque from the rotor changes. This allows the generator to accommodate variations in wind speed around the design speed and also helps reduce wear, particularly in the gearbox.

The main disadvantage of this type of generator is that it requires the grid to supply generator excitation and this creates a reactive power drain on the grid. In addition, the generator can only operate at or close to the grid frequency, so variable speed operation over a wide range is not possible. The induction generator is also less efficient than the main alternative. Besides that, modern grid codes require wind turbines to be able to support the grid rather than being supported by it, so this type of generator is not used in major utility wind turbines any more.

DOUBLY-FED INDUCTION GENERATORS

A more modern and more flexible version of the induction generator that is used in large wind turbines is a variant called the doubly-fed induction generator. In a conventional induction generator the generator stator is connected to directly to the grid and the rotor is a closed loop coil. However, in the doubly-fed induction generator the rotor has three phase windings and these are also connected to the grid supply through power electronic DC/AC converters. These allow a magnetic field to be generated in the rotor windings that interacts with the magnetic field in the stator windings, generating a torque. This torque depends on the strength of the two fields and the phase angle between them. By controlling this it is possible for the wind turbine to operate over a range of rotational speeds, roughly $\pm30\%$ of the grid frequency. A schematic of a doubly-fed induction generator is shown in Figure 5.2. Doubly-fed induction generators have been used in modern wind turbines with generating capacities of up to 5 MW.

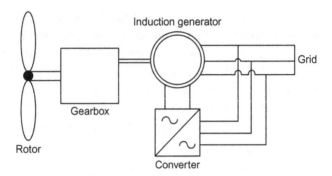

Figure 5.2 A schematic illustrating the operation of a doubly-fed induction generator. Source: Wikipedia commons https://en.wikipedia.org/wiki/File:Generator_principle_of_DFIG.jpg.

SYNCHRONOUS GENERATORS

The alternative to an induction generator is a synchronous generator. This produces an alternating current output that varies with the speed at which it rotates. In the most simple application of this type of generator to wind power, the turbine rotor is connected to the generator through a gearbox designed so that when the turbine is rotating at its design speed, the generator will be producing an alternating current output at the grid frequency. However, it requires that the generator and turbine be rotating at this frequency and no other, limiting flexibility. Small variations in rotor speed caused by wind speed variations and gusts are passed on to the grid as frequency fluctuations.

The synchronous generator has a rotating element, the rotor, which contains one or more magnets. These are most commonly electromagnets created by magnetic coils in the rotor which are excited to produce their magnetic fields by an externally provided direct current. This is normally taken by rectifying a small part of the output of the generator. An alternative to this which is becoming commonly used in wind turbine generators is to make the rotor using permanent magnets instead of magnetic coils. These are simpler as they require no excitation and are also lighter than the more conventional type of synchronous generator. However, they do require extremely powerful magnets containing up to 30% by weight of the rare earth element neodymium.

The synchronous generator suffers from the same problem as the conventional induction generator in that it can only operate at or very close to its synchronous speed. Otherwise it will feed power at the wrong frequency into the grid. The best way to overcome this is by using power electronics to convert the output of the generator from AC to DC and then back to AC at the grid frequency. With this arrangement the generator can operate at any speed and still feed power at the correct frequency into the grid. This is the basis for a variable speed generator.

The variable speed generator offers significant advantages. First it will allow the turbine rotor to turn at the optimum speed to match prevailing wind conditions. Then it also reduce fatigue stress of components because the rotor is allowed to change speed in response to gusty or turbulent conditions. Finally it also provides the wind turbine with the ability to provide grid support services for voltage and frequency control.

The generator in a variable speed drive train can be a conventional generator with coils in the rotor or it can use a permanent magnet generator. The latter are becoming more popular for large wind turbines. The generator will often be a low speed generator coupled to a simple gearbox and in some modern systems the gearbox and generator are integrated into a single unit. The combination of a hybrid generator/gearbox and low speed generator provides greater reliability by reducing the strain on the gearbox.

DIRECT DRIVE GENERATORS

The obvious extension of the use of low speed generators is to build a generator that can operate at the speed at which the rotor turns without the need for a gearbox. These direct drive generators are now common in many large wind turbines.

A direct drive generator operating at around 20 rpm will require many more poles than one turning at higher speed. For example at 125 rpm, 48 poles are required to generate a 50 Hz output. Such large numbers of poles are much more easily realized with permanent magnet generators. In addition, the direct drive generator must be larger than a generator designed to be coupled through a gearbox and the machine has to support the full torque from the rotor. With a gearbox, rotational speed is increased but torque is reduced.

In order to be as flexible as possible, direct drive generators also use power electronic AC/DC/AC converters in order to able to operate at

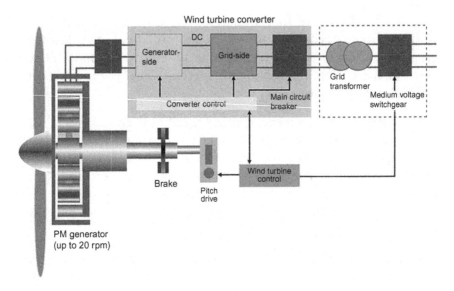

Figure 5.3 A permanent magnet direct drive wind turbine drive train and converter. Source: http://new.abb.com/ windpower/wind-power-generation/low-speed-full-converter-electrical-drivetrain-package. Graphic courtesy of ABB.

variable speed. This gives them the same flexibility discussed above in higher speed generators. A permanent magnet direct drive wind turbine drive train is shown in Figure 5.3. The elimination of the gearbox, combined with variable speed operation, has the potential to provide much greater reliability and this is particularly attractive for offshore wind turbine operators because offshore machines are much more difficult to maintain than those onshore.

Direct drive generators are used in wind turbines with generating capacities of 4–6 MW. For larger sizes, the sheer size of the generator may become a problem. This could eventually lead to a reversion to single stage gearbox/medium speed generators for very large machines.

CHAPTER 6

Nacelles and Towers

The nacelle of a wind turbine is the housing at the top of the tower which encloses the gearbox, generator, and other drive train components. It must be readily accessible so that maintenance can be carried out on the mechanical and electrical components and it should be weatherproof to protect these components. Some nacelles even contain viewing platforms.

The nacelle is mounted onto a large bearing that sits between the nacelle and the tower and allows the whole drive train and rotor assembly to turn relative to the tower below so that the wind turbine rotor is always facing into the wind. For optimum energy capture this position should always be maintained and the orientation is controlled by one or more yawing motors which rotate the nacelle and rotor as necessary.

The tower upon which the nacelle is mounted through the yaw bearing has to raise the turbine high enough from the ground or sea that the blades are clear of any turbulence caused by undergrowth, the local topography, or by heavy seas offshore. It has to be strong enough both to support the tower top structure and to resist the bending moment generated by the force of the wind on the turbine rotor, principally the result of aerodynamic drag, which is transmitted directly to the tower top. In addition it must be equipped with the means of access to the nacelle, either via ladders or a lift, and it must provide a conduit through which cables from the generator at the top of the tower travel to ground level. At ground (or sea) level the tower may also house a transformer that steps up the voltage from the generator to the distribution system voltage before connecting into the local electricity supply system, often via a dedicated substation if the turbine is part of a wind farm.

THE NACELLE

The nacelle of a wind turbine houses all the main components of the machine except the rotor. These include the drive shaft or

Wind Power Generation. DOI: http://dx.doi.org/10.1016/B978-0-12-804038-6.00006-2

shafts,[1] a gearbox if one is used, the generator, a brake to stop the turbine rotating in very low or very high winds, and a range of hydraulic and servo systems that control the blade pitch, the rotor speed, and the orientation of the complete nacelle structure by means of the yawing motors.

The tower top structures need to be as light as possible but even so a nacelle and its contents can weigh as much as 500 tonnes. To keep weight down, the nacelle housing is usually made from glass fiber reinforced resin (fiberglass) but this has to be carefully formed to retain strength over the relatively large panels which at times will be exposed to very high winds. These panels are mounted onto a rigid subframe to which the drive train is also bolted. This subframe may itself be massive in order to maintain the alignment of the drive train components in the face of the variety of distorting forces to which it is exposed during operation. Misalignment can cause high wear rates or failure of components, particularly the gearbox.

A nacelle may be equipped with a helicopter pad that allows maintenance crews to be lowered down to the tower top. This type of access can be particularly important offshore where seaborne maintenance of wind turbines can at times be difficult. There may also be a service crane to lift heavy equipment to the nacelle.

A nacelle and its contents will normally be assembled at the wind turbine manufacturer's factory, then shipped to the site and lifted into position at the top of the tower. As wind turbines have increased in size and nacelles have increased in weight, so the difficulty of lifting the nacelle into position has become greater. Specialist lifting equipment is now used when assembling the largest machines.

One of the key wind turbine manufacturers' design aims is to reduce the size and weight of the nacelle and its contents. Modern direct drive generators are helping to cut down both size and weight by removing the gearbox. This shortens the drive train, allowing for a more compact nacelle design. Inevitably, as turbine sizes grow, so will the size and weight of the nacelle.

---

[1]In a wind turbine with a gearbox the shaft from the rotor to the gearbox is called the low speed shaft while the shaft from the gearbox to the generator is called the high speed shaft.

TOWER-TOP REAL-TIME CONDITION MONITORING

Within the nacelle of a modern wind turbine there will also be found one further vital component, a real-time condition monitoring system to gauge the health of turbine components. Data from a range of sensors are collected and relayed to a control center where the data are stored and analyzed in order to predict future behavior of components as well as to be alert for any acute fault condition.

Factors that will be continuously monitored include the power output of the wind turbine, the reactive power and the power factor. Other sensors will measure wind speed, gearbox and bearing oil temperatures, generator winding temperature, and the nacelle temperature. There may be acoustic sensors that can measure low frequency vibrations that might indicate problems with the tower or foundation while other acoustic sensors may detect changes in the blades, such as the roughness of the blade surfaces, leading to greater turbulence and more fatigue loading. Measuring the rotational speed and the torque generated by the rotor may allow any imbalance in the rotor to be highlighted. By combining and analyzing all this sensor information in real time, it becomes possible to see when components are wearing more than expected or where a fault is beginning to develop. Operators can then modify the mode of operation of the turbine to protect it from failure until a maintenance team can be brought in to correct the fault. This type of condition monitoring is also used to build a maintenance schedule for each machine. Real time condition monitoring was developed first for large offshore machines that are difficult to reach and service. However the technology is now used in both onshore and offshore fleets of wind turbines.

THE YAW BEARING AND MOTOR

The yaw bearing sits between the nacelle and the tower and provides the means to rotate the tower-top structure to keep the rotor facing into the wind. In this position the yaw will be subject to all the drag and twisting forces on the turbine/nacelle combination and it must be strong enough to transmit these to the tower, which is the ultimate guarantor of wind turbine stability, while being itself unaffected by them. It must do this while resisting corrosion and wear so that it remains in operation for the lifetime of the turbine. Design life is usually 20 years or more.

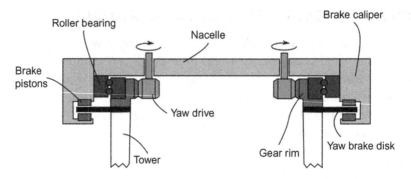

Figure 6.1 A roller yaw bearing for a wind turbine. Source: Wikipedia commons https://en.wikipedia.org/wiki/ File:Roller.yaw.bearing.svg.

There are two principle types of yaw bearing in use in modern wind turbines, a roller bearing and a glide bearing. The roller bearing operates on the same principle as logs laid down on the ground in order to move a heavy load by rolling it across them. It offers the smoothest low friction bearing but turbines with this type of bearing have to be equipped with brakes because of the ease of movement. A roller yaw bearing for a wind turbine is shown in Figure 6.1. A glide bearing uses a low friction material as the gliding surface upon which a steel element can rotate. Friction is higher than in a roller bearing and wear is normally of greater concern in this type of bearing. Most large wind turbine manufacturers appear to prefer the roller bearing.

In addition to the bearing, the yawing system must be equipped with a series of motors that can orient the nacelle relative to the tower. There are usually several of these positioned around the perimeter of the bearing, all controlled by computer systems and wind sensors within and on top of the nacelle.

Cables from the generator housed within the nacelle carry power through the yaw bearing to the base of the tower. These heavy duty cables must be capable of carrying up to 4000 A from a 3 MW generator with a 690 V output. As the nacelle turns on the yaw bearing, so these cables will inevitably become twisted. The computer systems in the nacelle must keep track of this twisting. In a typical large wind turbine the nacelle can complete three revolutions before the computer will rotate the yaw bearing backwards to unwind the cables.

WIND TURBINE TOWERS

The tower of a horizontal axis wind turbine provides the support upon which the turbine electromechanical components are held in the correct position to capture the most energy from the wind. For both onshore and offshore wind turbines, the tower must raise the rotor so that its blades are both clear of the ground and clear of any obstacles. It must also place the rotor hub at such a height that the tips of the blades are clear of any turbulent layer of air close to the ground as this can cause additional fatigue loading as well as reducing energy capture. The higher the turbine rotor is mounted at any site, the stronger the wind, so it is better to raise the nacelle on the highest tower possible. In some cases turbine developers will choose to place a rotor on a higher tower than usual if the gain in power is sufficient at the higher elevation.

While there is no rule, the tower height is usually somewhere between two and three times the blade length. This height provides the optimum balance between energy capture and cost. However, it will always be site sensitive. The height of a wind turbine is usually defined by the hub height which is the height at which the center of the rotor sits above ground or sea level.

The typical height for a 3 MW wind turbine tower is around 80 m, although hub height can stretch to 100 m for some turbines of this size. A commercial 7.6 MW turbine has a hub height of 135 m, while a project to test the feasibility of a 20 MW wind turbine proposed a hub height or around 150 m. Such large machines as this are only ever likely to be used offshore.

Wind turbine towers have been constructed in a variety of ways. Many early turbines used lattice steel structures. Lattice towers are relatively cheap to build and they produce little wind shadow. A wind turbine with a lattice steel tower is shown in Figure 6.2. However, they are visually unappealing and they have high maintenance costs. As turbine size grew lattice towers were considered to be less practical and when towers extended above 60 m tubular towers began to grow in popularity. Attempts have been made to construct these from concrete but that proved costly and in most countries the industry gravitated towards tubular steel towers. These towers, which are conical in form, have become the standard means of construction for tower heights of up to 80 m. A wind turbine with a tubular steel tower is shown in Figure 6.3.

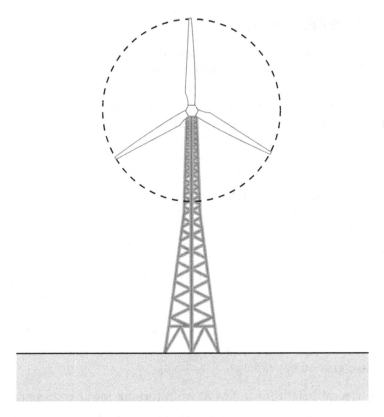

Figure 6.2 A wind turbine with a lattice steel tower. Source: World Steel Association. The image is taken from http://www.worldsteel.org/dms/internetDocumentList/bookshop/worldsteel-wind-turbines-web/document/Steel%20solutions%20in%20the%20green%20economy:%20Wind%20turbines.pdf.

A tubular steel tower for an 80 m hub height will have a diameter at the base of around 4.5 m at the base and 2 m at the top. The tower will be constructed in three or perhaps four sections of 20–30 m in length and these are then bolted together at the site. The sections themselves are made from sheet steel that is cut, rolled, and welded to produce the required shape. Turbine towers of 80 m can weigh up to 230 tonnes making them both costly and difficult to transport and to erect.

Increasing the height beyond 80 m means building a much stronger tower because the forces acting upon it are greater. To maintain the same design it is necessary to add a significantly greater mass of steel. Depending on commodity costs, this can make tubular steel towers of 100 m and more uneconomical. In addition the base diameter can exceed 5 m and this can limit the ability to transport the section by road.

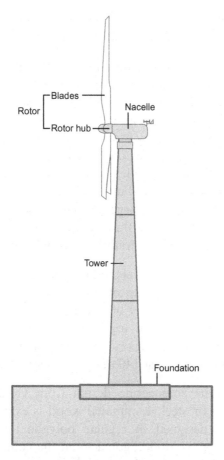

Figure 6.3 A wind turbine with a sectional tubular steel tower. Source: World Steel Organisation. The image is taken from http://www.worldsteel.org/dms/internetDocumentList/bookshop/worldsteel-wind-turbines-web/document/Steel%20solutions%20in%20the%20green%20economy:%20Wind%20turbines.pdf.

Concrete has been used for towers of 80 m and less in countries where concrete itself is cheap. Beyond 80 m, concrete can be more economical elsewhere too. A concrete tower can be made up from in a large number of precast sections that are assembled and joined at the site. These sections can be smaller than the sections of a tubular steel tower but against this a concrete tower will generally be much heavier than a steel tower. Some manufacturers have experimented with hybrid concrete/ tubular steel towers, particularly during times when iron and steel prices have been high. These use concrete sections for the base onto which a conventional tubular steel tower is bolted. A 3.2 MW wind turbine with 114 m blades has been erected in Germany on a hybrid tower of this type which is 143 m high. These towers are complicated to assemble but offer a more economic solution than the alternative tubular steel.

Hybrid towers may be one solution to tall towers, but as tower sizes creep beyond 100 m, so designers have begun to look for alternative ways of constructing them that are both cheaper and make transportation and assembly easier. A programme in the United States funded by the U.S. Department of Energy examined the use of composite materials in the design of a scalable tower capable of exceeding 100 m in height. The research highlighted that the composite materials examined, although lightweight, could not achieve the required level of stiffness and would be exposed to large resonance oscillations. Instead the programmed developed a space frame tower that proved to be cheaper and easier to assemble than the traditional tubular steel tower.

An alternative approach, developed in Germany, uses a bolted steel structure. The tower is made up of several vertical sections but each section is fabricated as a number of steel shells that can be transported to the site before being bolted together to create each section. This highly modular approach makes transportation much easier and should allow towers of at least 140 m to be constructed.

Meanwhile another pioneering design from Germany uses wood instead of concrete or steel. Laminated wood is extremely strong and light. Towers are designed in regular polygonal shapes—typically, hexagonal, octagonal, or dodecagonal—and the laminated sections are glued together at the site. Developers claim that towers of up to 200 m can be constructed in this way.

WIND TURBINE FOUNDATIONS

The complete wind turbine structure, including the nacelle and tower, must be anchored securely at ground level if it is to be able to perform its role without falling over. To achieve this, the tower needs to be attached to a turbine foundation. Turbine foundations have to be extremely substantial if they are to carry out their role successfully.

The most common type of foundation used for onshore wind turbines is a gravity foundation (sometimes called a spreading foot foundation) which relies primarily on the mass of the foundation and the soil into which it is placed to hold the tower in place. This type of foundation will have a relatively large area compared to its depth. The foundation will comprise a massive reinforced concrete structure fabricated in a hole excavated at the site of the wind turbine. It will be proportioned initially so that it is stable when subjected to the over-turning forces from the turbine and tower. This will depend on the mass and foot area of the foundation as well as on the additional resistance conferred by the soil in which it is buried.

Other properties of the foundation will depend on the geotechnical properties of the soil such as its bearing capacity and deformability. The foundation must resist horizontal movement due to the forces on the turbine and tower. It must also resist rotational forces. In both cases this will depend on the properties of the soil. Differential settlement must be taken into account, as must durability. And finally the cost of the foundation will be an important factor, particularly with large wind turbines.

Where the soil lacks strength or where there is no soil layer, other forms of foundation can be used. For rock substrate, the most common type of foundation will be a rock socket whereby a hole is bored into the rock and the foundation is laid within it. The rock beneath this type of socket must be able to bear the load of the tower and nacelle and it must also be able to resist the lateral forces that will result from the wind's action on the turbine. If the bedrock is at a very shallow depth it may be possible to create a foundation by essentially bolting the tower to steel components fixed to the rock. Meanwhile if the soil at the surface is not capable of supporting the load generated by the wind turbine but substrata can, then piles can be sunk to that level. The load-bearing substrata will then support the weight of the wind turbine transmitted through the piles while lateral forces will be resisted by the earth around the deep piles.

WIND TURBINE RESONANCES

Wind turbines are complex structure with major rotational components and these have the potential to excite resonant frequencies within the turbine structure. One of the most important components in which a resonance can be excited is the turbine tower which as noted above is usually made of tubular steel, a material that is potentially flexible and capable of vibrating in response to stimulation. To combat this, towers must be sufficiently stiff to resist resonant vibrations.

Vibration can be caused by the rotation of the turbine and this can be a particular problem in the case of variable speed wind turbines because these operate over a range of rotational speeds and can sometimes excite the tower resonance at several frequencies which are harmonics of the fundamental frequency. In addition to the rotational frequency of the turbine rotor, there is another frequency called the crossing frequency which is the frequency at which the blades cross the tower, generally three times the rotor rpm. When a rotor crosses the tower the latter is momentarily shielded from the wind and this can set up a sympathetic oscillation. Wind turbine design has to take these potential resonances into account, using methods to avoid the excitation of tower frequencies and applying damping techniques to reduce the magnitude of any vibrations that are excited.

If vibrations are not controlled, they can lead to additional fatigue loading of the turbine components and this can cause early failure. In the worst case an uncontrolled resonance is capable of causing catastrophic failure.

CHAPTER 7

Wind Farms, Electrical Optimization, and Repowering

A wind farm is a collection of wind turbines that operate together so that from the perspective of the grid they appear to form a single power station. While single turbine installations can be found in most countries and regions, by far the most important grid-connected wind turbine capacity is located in wind farms. A wind farm may be as few as two or three turbines or it may run to hundreds.

The largest in the world is the Gansu Wind Farm Project in China which, in 2015, had a nominal generating capacity of 7900 MW. Although a single project, the scheme comprises multiple wind farms. For example, the first phase involved eighteen 200 MW wind farms and two 100 MW wind farms. The project is expected to grow to 20,000 MW of installed capacity by 2020 and involves thousands of wind turbines. Although the total has not been published, at a nominal 3 MW for each turbine the number must exceed 2600 individual turbines.

While the Chinese wind farm is the largest combined project, the largest single array wind farm is probably the Alta Wind Energy Center in California which will have a generating capacity of 1550 MW from 600 wind turbines when complete. There are seven other wind farms in the United States with capacities of 500 MW or above, five of them in Texas. Meanwhile the largest onshore wind farm in Europe is believed to be the Fantanele-Cogealac Wind Farm in Romania with a capacity of 600 MW from 240 individual wind turbines. The largest offshore wind farm currently in operation is the London Array with 630 MW of generating capacity from 175 wind turbines. A schematic of an offshore wind farm is shown in Figure 7.1.

Building a wind farm with multiple turbines involves a wide range of considerations that do not come into play for single turbine installations. For optimum energy capture it is important to pay attention to the

Wind Power Generation. DOI: http://dx.doi.org/10.1016/B978-0-12-804038-6.00007-4

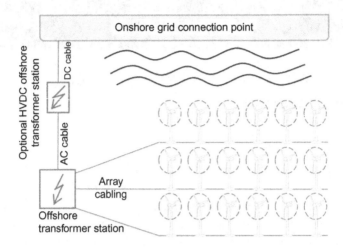

Figure 7.1 Schematic of an offshore wind farm. Source: E.ON Offshore Wind Energy Handbook http://www. eon.com/content/dam/eon-com/en/downloads/e/EON_Offshore_Wind_Factbook_en_December_2011.pdf.

pattern in which the turbines are laid out. Interference must be minimized if energy capture is to be maximized. Cabling layout is important too, particularly where hundreds of turbines are being connected together. Wind farms are normally provided with a dedicated substation and power from each turbine is fed to that substation across a complex web of cables. If the cable layout is not optimized, significant losses can result.

ARRAY OPTIMIZATION

The layout of a wind farm can have a major effect on the electrical output it produces and therefore ultimately on its economic viability. When a single wind turbine is installed at a wind site, wind conditions at the site will be the main factor determining the type and size of turbine and its location. However when multiple wind turbines are to be installed then the interactions between the turbines become a vital consideration too.

Like a large jet airliner, a wind turbine will leave a trail of air turbulence in its wake. This turbulence will affect anything following behind so that in the case of aircraft, gaps have to be left between take-off slots to allow turbulence along the runway to dissipate. A wind turbine does not move but instead leaves a wake downwind of its location. If another turbine is placed too close to the first it will experience the turbulence in this wake and its energy capture and output will be reduced. This situation is complicated by the fact that the wind direction varies from day to day and this must be accounted for, too, when planning the wind farm layout.

Energy losses resulting from poor turbine layout can amount to as much as 10% of a wind farm's predicted output. On the other hand good optimization can actually allow more energy to be captured with one offshore project achieving 15% more energy output that projected at the outset as a result of array optimization.

In order to improve the understanding of the wind and the layout of wind farms, developers are using a range of tools including fluid dynamic modeling of the wind farm site and satellite and ground based measurements of the wind speed and direction to build up a picture of the wind conditions at the site. Key data will include the wind magnitude and directional distribution and the seasonal fluctuations. Site turbulence needs to be understood and then when a good set of wind data is available, analysis can be carried out to match the best turbines to the site. Some sites might be more productive when equipped with a smaller number of very large turbines, others will benefit from smaller turbines. Tower height will also be a factor. Then, once a turbine has been selected, its performance at the site must be modeled in different arrays in order to find the best solution. Electrical layout and structural engineering will also play its part in determining how the wind farm should be laid out, and economic analysis will play an important role too.

An EU funded project, TOPFARM (Topology OPtimization of a wind FARM) set out to generate a tool that could be used to optimize wind farm layout from the developer's perspective in order to maximize the economic viability of a project. This used computer modeling to look at turbine wake effects on both power production and turbine component fatigue. While this project and its successors were aimed at offshore wind farms, many of the same principles can be applied onshore.[1]

Another approach to array structure is the idea of using wind turbines of different sizes within a single array. While most wind farms are built using a single wind turbine, varying the size can potentially have a positive effect on the level of turbulence and therefore the turbine fatigue

---

[1]Wind Farms Design and Optimization, Pierre Elouan Réthoré, DTU Wind Energy.

loading. Modeling the effects of turbulence over the lifetime of a wind farm can be used to predict exposure to fatigue loading and consequent component wear. This can be used to compute the lifetime maintenance costs which are fed back into the economic model of the wind farm. Advanced modeling techniques of this type will normally be used when any new wind farm is being developed.

CABLE LAYOUT

Another aspect of wind farm design is the cable layout. In a wind farm, whatever its size, the power from each turbine will have to be carried to a transformer substation before delivery to the grid. A wind turbine generator will produce an output voltage that is typically between 450 and 690 V although design output voltages vary. This voltage will then be stepped up to a distribution system voltage of perhaps 35 kV by a dedicated transformer at each turbine. The output from this step-up transformer is transmitted to the wind farm substation where power from all the turbines is collected and stepped up again to the transmission system voltage of 130 kV or more.

The cabling from each turbine may therefore carry a current of up to 100 A when the turbine is generating at full output. Over a long cable run, this current can lead to a significant resistive energy loss, generating heat that is dissipated to the local environment. In order to limit this loss the length of cabling from each turbine transformer to the wind farm substation needs to be optimized. When there may be hundreds of turbines to be connected, this will involve complex modeling so that a cabling array is developed that leads to the lowest overall losses.

As with array optimization, economics form a part of this equation. Losses can usually be reduced by using higher quality cable but this will cost more and so affect the overall project viability. Another factor affecting losses is cable temperature which, in turn, depends on the current within the cable. Wind farms will generally use a radial cabling array, with the substation as close to the center as possible and with the largest current flows through the shortest cable runs. However this still leaves a range of possible layouts that can only be filtered using computer modeling.

WIND TURBINE TRANSFORMERS

When wind turbines and wind farms are being designed it is easy to overlook the turbine transformer yet this component plays an important role in overall wind farm efficiency. Turbine and wind farm designers have, naturally, looked for the most economical option when choosing a transformer for their wind turbines and this has usually been a standard, off-the-shelf component. These standard transformers are typically designed for steady state operation or daily cycling whereas the duty cycle of a wind turbine is never steady-state and may cycle between full and very low power several times in a day. Standard transformers are often unable to cope with this duty cycle and consequently there have been numerous instances of wind turbine transformers failing. Anecdotal evidence suggests that wind farm operators regularly keep a supply of replacement transformers in case of such failures. Regular failures and replacements mean that the savings associated with the choice of a low-cost transformer are soon swamped by maintenance costs and loss of revenue as the result of unit outages. This has led manufacturers to acknowledge that a wind turbine needs a transformer that has been designed with this specific role in mind.

The unusual duty cycle of a wind turbine can have several consequences for transformer design. A wind turbine in a typical onshore wind farm may have a load factor of 35%. This will involve a variety of loads through the transformer, leading to continuous thermal cycling. This type of cycling can induce stress in all of the transformer components including windings, gaskets, and seals. Meanwhile frequent cycling can also cause problems in the dielectric oil in a transformer that will over time damage the insulation.

Another problem that arises with low transformer loading is that losses in the transformer core become amplified relative to coil losses, which would be considered the most important at higher loads. Meanwhile many wind turbine transformers are fed from the wind turbine itself through an electronic inverter so that the turbine can operate at variable speed. However inverters produce spikes and harmonics as well as the 50 Hz or 60 Hz output they are designed to deliver and these spurious signals can produce heating and losses and damage components such as the transformer insulation.

For all these reasons transformer manufacturers are now offering transformers that have been specially optimized for wind turbine operation. While these tend to be more expensive than the standard units they are replacing, their greater efficiency and reliability can offset this, providing an economic advantage.

REPOWERING WIND FARMS

The life of a wind farm is generally considered to be around 20 years. What happens when this time is up? In all probability the wind turbines will still be operational but at this stage in their lives they will be operating at lower efficiency than when they were first installed. Their reliability will have fallen too. Meanwhile in the 20 years since the wind farm was developed and these turbines were installed wind turbine technology will probably have advanced significantly.

One option is simply to decommission the wind farm if it is no longer commercially viable, take down the turbines and return the land to its previous state. However gaining permission to build a wind farm is usually a difficult and lengthy process. Therefore it often makes poor commercial sense to abandon as site which has approval for a wind farm to operate. Instead, it will often be economically attractive to replace the old turbines with new units that provide higher output, often from far fewer machines.

This process, known as repowering, has become an important part of wind energy development as more wind developments reach the end of their lives. It can also be an option before that time is reached, particularly when wind technology is advancing rapidly so that turbines designed and installed perhaps 10 years previously have become obsolete and have been superseded by bigger, more efficient machines. However there is another side to repowering. These 10 year old machines are still viable and efficient wind turbines and when they are taken down, they can be sold for relatively modest prices. This has allowed wind to be developed in other parts of the world where the cost of new wind turbines would be prohibitive.

While the ability to repower at an existing wind farm is unlikely to be a formality, it can be economically extremely attractive as well as providing and environment improvement. For example, a small wind farm built in the UK's Lake District in 1993 comprised 12 wind turbines, each of 400 kW, providing a total installed capacity of 4.8 MW. The owners of the wind farm proposed replacing the existing turbines with six new ones each of 2 MW or 3 MW, for a new installed capacity of 12–18 MW. Of course increasing the output of a wind farm by two to three times can present interconnection problems if the existing grid connection cannot support the new output. Even so, the economic advantages of such projects have been widely recognized.

The greatest potential for repowering is to be found in countries that were the pioneers of wind development such as the USA, Germany, and Denmark. However, there are many other countries such as the UK that now have wind farms which are 20 or more years old. Meanwhile the availability of large numbers of second-hand wind turbines can make it easier to obtain wind turbines when the lead in time for the manufacture of new units is long.

Another trend that is allied to repowering is the refurbishment and redesign of existing turbines so that their lives can be extended. This can allow existing turbines to continue to operate at a site where repowering with fewer, larger turbines is not possible. It also offers a way of adding value to the growing number of second hand turbines that are available in what is becoming a global market. While this type of refurbishment has been pioneered by small independent companies, it is now being picked up by some major wind turbine manufacturers too.

CHAPTER 8

Small Wind Turbines

Small wind power is an important but sometimes overlooked part of the wind turbine market. According to the American Wind Energy Association (AWEA), small wind refers to wind turbines with generating capacities of 100 kW or smaller. In other parts of the world a 100 kW wind turbine is considered a large machine and the limit for small wind turbines is set at 10 kW. Sizes may be as small as 1 kW, suitable for off-grid generation for a single dwelling. A 10 kW turbine could supply power to a commercial enterprise. At the upper end of the AWEA small wind power range, a 100 kW machine would be capable of supplying power to a small community.

While there is a good deal of overlap, wind turbines in the small wind category often use different technologies to their larger cousins. While multimegawatt wind turbines follow a standard pattern today, it is still possible to find a wide range of small wind turbine designs including many vertical axis wind turbines. The number of blades on each rotor, different tower or mast designs, and alternative types of generator all contribute to the small wind turbine diversity.

Putting a figure on the size of the global installed capacity of small wind turbines is extremely difficult because of a dearth of data. The AWEA, which has carried out the most comprehensive recent studies of small wind power suggested in a 2009 report that the United States accounted for around half of the global market for small wind turbines. At the end of 2011 the cumulative installed capacity of small wind turbines in the United States was just under 200 MW.[1] While it is naive to extrapolate from this data without further information about the world market, there is no other way of gauging the global total; taking the naive route, we can guess that the global small wind installed capacity is perhaps twice this figure, or 400 MW. This can be

---

[1] AWEA Small Wind Turbine Global Market Study: Year Ending 2008, American Wind Energy Association, 2009. 2011 US Small Wind Turbine Market Report; Year Ending 2011, American Wind Energy Association 2012.

Wind Power Generation. DOI: http://dx.doi.org/10.1016/B978-0-12-804038-6.00008-6

compared with the total global wind capacity in 2011 of 238,435 MW. On the other hand, the number of small wind turbines making up the United States total in 2011 was 151,300, suggesting by the same logic as above that there are a global total of 300,000. The estimated total number of commercial large wind turbines in operation in 2014 has been put at 250,000, so there may be more small wind turbines than there are large ones.

SMALL WIND TURBINE MANUFACTURERS AND APPLICATIONS

Based on the AWEA global survey, there were at least 219 companies that manufactured or planned to manufacture small wind turbines in 2008. While the largest number (66) was in the United States, there were significant numbers in Japan (28), Canada (23), the UK (18), Germany (16), and China (14). Of the remainder, most were in Europe but there were five were in Africa, three in the Middle East, and three in India. While some small wind turbines are exported, mostly in the larger categories, many will be sold into domestic markets. The breakdown of manufacturers can therefore be used as a coarse guide to the main areas in which small wind turbines are to be found. The AWEA figure for the number of manufacturers is corroborated by a small wind turbine website which lists 401 small wind turbines from 229 manufacturers across the world.[2]

Applications for small wind turbines are diverse but the main market sectors appear to be the domestic and commercial markets. In the developing world this probably also includes small industrial units, particularly where grid connections are not available. The market can also be broken down into grid connected and off grid units. Based on the annual capacity installed, globally, in 2008, the largest sector is the grid connected small wind sector. However, when the numbers of units are examined, by far the largest proportion, 74%, sold in 2008 globally were off-grid.

This breakdown indicates that off-grid small wind turbines are often smaller units. In the United States in 2011 only 3.4% of units sold were larger than 10 kW. However these units accounted for 61% of the total capacity sold. Larger units will almost always be grid connected and may sell power to the local grid as well as supplying electricity to the owner of the wind turbine. However a significant proportion of these

[2]www.allsmallwindturbines.com

larger units are also used in wind-diesel hybrid systems to provide reliable power in remote locations. Such systems may be combined with solar power and battery storage. Small, off-grid units will normally be used where there is no grid connection of where the wind power is all to be used by the owner. In many cases these will also be connected to a battery storage system so that continuous power is available irrespective of whether the wind is blowing.

SMALL WIND TURBINE TECHNOLOGIES

Small wind turbines encompass a wide range of technologies. Most are horizontal axis machines but there are also a variety of vertical axis small wind turbines available. At the lower end of the scale it is possible to buy all the components needed from different suppliers and produce a homemade wind turbine. However, most turbines, even at the small end of the scale will be supplied complete and ready to connect either to the grid or to a local distribution system. Meanwhile small wind turbines in the 10–100 kW range will often be essentially smaller versions of the large commercial machines used by utilities. Figure 8.1 illustrates a typical small wind turbine system.

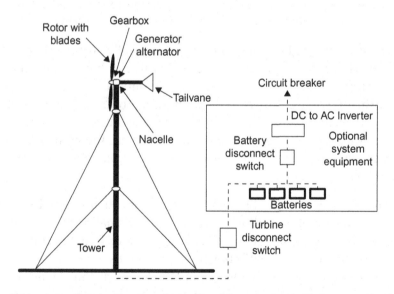

Figure 8.1 A typical small wind turbine system. Source: Natural Resources Canada. The image was taken from this web page http://www.omafra.gov.on.ca/english/engineer/facts/03-047.htm. Courtesy of "Stand-Alone Wind Energy Systems – A Buyer's Guide." Natural Resources Canada, 2003. Reproduced with the permission of the Minister of Natural Resources Canada, 2015.

Like large wind turbines, most larger small wind turbines use three bladed rotors since these represent the optimum compromise between cost and efficiency. These will have an aerofoil cross-section and may be made from fiber-reinforced resin since this offers a cheap, lightweight method of manufacture. Commercial small wind turbines will be provided with an integrated turbine-generator assembly, often without a gearbox since a small wind turbine rotor can turn sufficiently fast to drive a generator without a gearbox. However, many of these small wind turbine systems will also be equipped with an AC/DC converter to convert the output from the generator to direct current suitable to charge a battery.

There are many alternatives to this conventional small horizontal axis wind turbine. Some have curved aerofoil blades and there are machines with two blades as well as with four or more. There are also a range of small vertical axis wind turbines. These include Darrieus wind turbines, discussed in Chapter 3 and Savonius wind turbines. The latter are very simple drag-type devices with a vertical rotor that in horizontal cross-section is shaped like the letter S, with two scoops that are driven round by the wind. The principle upon which they work is similar to that of an anemometer, with cup-shaped rotor blades, used to measure wind speed. Other small vertical axis wind turbines use novel rotor designs unique to individual companies.

Small wind turbine manufacturers generally try to keep their designs as simple as possible since they will not have utility maintenance staff to keep them running. One simplification is to use passive yaw control, relying on a vane which projects from the opposite end of the main shaft axis to the rotor to keep the turbine oriented either upwind or downwind, depending upon the design. Integrating the rotor directly with the generator helps reduce the component count to a minimum.

Another area for simplification is the tower, often referred to as the mast in the case of a small wind turbine. Some small wind turbines use tubular steel masts, similar in concept to those of large machines, that are strong enough and stiff enough to provide support for the turbine and generator assembly. However many use much thinner masts that are held in position using guys to resist bending forces. Others use space frames or tripods and there are a large number of traditional

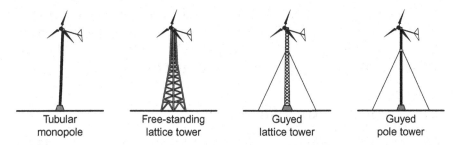

| Tubular | Free-standing | Guyed | Guyed |
| monopole | lattice tower | lattice tower | pole tower |

Figure 8.2 Types of towers for small wind turbines. Source: Sustainability Victoria (the state of Victoria, Australia). The image was taken from the document 'Victorian Consumer Guide to Small Wind Turbine Generation'. It can be found here: http://www.sustainability.vic.gov.au/∼/media/resources/documents/Publications %20and%20Research/Knowledge%20archive/Small%20scale%20renewable%20technology/Wind%20in%20Urban% 20Areas/Archive%20Small%20Wind%20Generation%20Jul%202010.pdf.

lattice towers in use with small wind turbines. In some cases wind turbines are roof mounted, so requiring shorter masts than they would if they were ground mounted. Figure 8.2 shows a variety of small wind turbine towers and supports.

Generators for small wind turbines vary widely. The largest machines may be equipped with generators that can produce grid frequency electricity. However many of the smaller units will have simple generators that are incapable of providing synchronous power. A common generator for some grid connected small wind turbines is the induction or asynchronous generator that was used widely during the early days of commercial wind generation. These are essentially electric motors that are used as generators.

An alternative to this type of generator is the permanent magnet direct current (PMDC) generator. Again these are usually motors, driven from a direct current supply rather than an alternating current supply but if they are rotated at a speed greater than the design rotation speed, then they will generate direct current. A direct current generator is often called a dynamo and there are dynamos designed for use in heavy duty vehicles such as trucks or buses that can be used for wind generation too. Other designs have used a rim generator, where the rotor permanent magnets are around the perimeter of the wind turbine rotor and this is then enclosed by the stator windings. For small wind applications the DC output of a PMDC generator will match well with the need to be able to charge batteries in order to store wind energy for use when the wind does not blow.

When a small wind turbine is to be grid connected then the generator and interconnection will have to be much more sophisticated. In this case some form of solid state DC/AC converter, or AC/DC/AC converter will be required to comply with grid codes.

SITING SMALL WIND TURBINES

The rules for siting small wind turbines are exactly the same as for large wind turbines. This means that the average wind speed and the wind speed distribution should be known. For small wind generation, a minimum wind speed of around 3.5 m/s is usually required although some designs claim to be able to operate with as little as 1 m/s. However, energy capture at this speed is going to be very low. In an urban environment a wind speed of less than 5 m/s is probably the economic limit. Maximum wind speed is generally around 12 m/s.

Gaining access to this type of wind regime will depend on the location and on the mast height. While many small wind turbines have masts of around 10 m, this will often not be sufficient to access a good, smooth wind regime. If the turbine is too low it will be affected by ground-generated turbulence. Turbulent air is both less efficient for generating wind power and also leads to high levels of fatigue stress which can lead to early turbine failure.

Ground generated turbulence is a problem for all wind generation but it is particularly problematic in urban areas where buildings will dramatically affect the air flow. When planning the siting of a wind turbine the roughness of the local terrain is important. Roughness can be classified on a scale of 0−4, with 0 being the roughness of the sea and 4 defines a typical urban landscape with trees and buildings. Building-mounted wind turbines in urban areas need to be raised above the tops of the highest buildings in the local region if they are to have any change of operating efficiently. Mast mounted turbines will also need to raised higher than they might be in a rural environment. For a single building, not surrounded by other, higher buildings, the turbine will probably need to be 10 m above the highest point on the roof.

Small wind turbines are generally less efficient than large wind turbines. A typical rotor may be able to capture 35% of the wind energy, compared to up to 50% for a large machine. In addition a small wind turbine will suffer from yawing losses when gusts cause the turbine to move out of its optimum orientation.

The visual impact of a small wind turbine is also likely to be important when planning a small wind installation. In urban areas it can be difficult to find suitable sites for horizontal axis wind turbines that do not intrude visually. However vertical axis wind turbines can offer more interesting designs that are often easier to integrate visually. Visibility will be less of an issue in rural environments but it must still be taken into account. Factors that can aid integration include the color of the turbine and its height relative to other features in the surroundings such as buildings or electricity pylons.

CHAPTER 9

Offshore Wind

The amount of energy available from offshore sites is considered to be much greater than that available onshore. The wind regime is generally better at sea too because there are no hills or mountains to interrupt its flow. This means wind speeds are generally higher, the wind is more reliable, and there is less turbulence. In consequence more energy can be captured with smaller numbers of wind turbines. Estimates of the total amount of energy available offshore are tentative at best but according to the Global Wind Energy Council the energy available from European offshore wind energy sites could provide seven times the current energy demand while in the United States it could provide four times the energy demand.

Another advantage of offshore wind is that it is often much easier to gain planning permission for an offshore wind farm than it is for an onshore facility. With many of the world's largest cities located at coastal sites, wind farms off the coast adjacent to these cities could potentially provide the large demand centers with locally-generated renewable power. Against that, building wind farms offshore is much more challenging than building onshore and the costs are much higher.

The modern offshore wind movement can be dated to the late 1980s and early 1990s with a number of small European pilot projects. Development was slow during the decade of the 1990s so that by 2000 there were only 45 MW of offshore generating capacity globally, all in European waters. Since then the installed capacity has accelerated, as shown in Table 9.1. In 2010 there were 3087 MW and by 2014 the cumulative capacity reached 8759 MW of wind generating capacity at offshore sites. This represented around 2% of the total global installed wind capacity. However the Global Wind Energy Council has projected that by the end of the second decade of the twenty-first century it will reach around 10% of global wind capacity.

Wind Power Generation. DOI: http://dx.doi.org/10.1016/B978-0-12-804038-6.00009-8

Table 9.1 Annual Global Offshore Wind Capacity and Additions[1,2]		
Year	Capacity Added (MW)	Cumulative Installed Capacity (MW)
2000	14	45
2001	50	95
2002	160	255
2003	259	514
2004	66	580
2005	90	670
2006	211	881
2007	242	1123
2008	350	1473
2009	583	2056
2010	1031	3087
2011	1624	4711
2012	704	5415
2013	1671	7086
2014	1673	8759
Source: *Business Insights, Global Wind Energy Council.*		

The offshore wind industry has remained essentially a European industry with virtually all the offshore capacity located in the North Sea, Irish Sea, the Baltic Sea, and the English Channel. The leading offshore wind nation is the United Kingdom which had 4494 MW installed in its waters at the end of 2014, as shown in Table 9.2. Denmark had 1271 MW and Germany 1049 MW. Other important offshore wind nations include Belgium, the Netherlands, and Sweden. Development of offshore wind capacity in other parts of the world has been slow, primarily as a result of the cost of building offshore. The United States, has been notably reluctant to build, despite having considerable offshore wind potential along its Eastern and Western Seaboards where some of the nation's large cities are located. At the end of 2014 there was a single 20 kW offshore wind turbine in waters off Maine. In Asia, there are two demonstration projects in China with a total generating capacity of 658 MW and 5 MW in South Korea. Japan was also slow to adopt the technology but the shock of the Fukushima nuclear disaster in 2011

[1]The Future of Offshore Wind Power Generation, Paul Breeze, Business Insights, 2011.
[2]Figures are taken from the Global Wind Energy Council website.

Table 9.2 Distribution of Offshore Wind Capacity in 2014 by Country[3]	
Country	Installed Capacity (MW)
United Kingdom	4494
Denmark	1271
Germany	1049
Belgium	713
China	658
Netherlands	247
Sweden	212
Finland	26
Japan	50
Finland	26
Ireland	25
South Korea	5
Spain	5
Portugal	2
USA	(0.02)
Source: *Global Wind Energy Council.*	

has forced the country to look at alternative sources of energy of which offshore wind is considered a valuable component. There are now 50 MW of offshore wind capacity in Japanese waters.

OFFSHORE WIND TURBINE TECHNOLOGY

Offshore wind turbines are similar to onshore wind turbines and use substantially the same technology. The only significant difference as far as energy capture is concerned is that they are often larger. As with onshore machines, those used offshore are horizontal axis wind turbines with three-bladed rotors. Drive-trains for offshore wind turbines are identical to those used on land too; some use gearboxes and relatively high speed generators while others are equipped with direct drive between the turbine rotor and the generator. Towers are of similar construction. Offshore wind turbine towers are generally built from tubular steel sections. However, foundations for offshore wind turbines are substantially different to those used onshore.

[3]Figures are taken from the Global Wind Energy Council website.

Aside from that, the main difference between onshore and offshore wind development is the environment in which the wind turbines must operate. The offshore wind regime is generally more fierce than onshore with the potential for much higher average wind speeds and the conditions at sea are much more challenging too due to the corrosive nature of salt water. This means that offshore wind turbines have to be more rugged than similar onshore machines. To combat this, marine technologies used to prevent seawater damage to offshore oil and gas installations have been adapted for use by the wind turbine industry. In addition, the difficulty of carrying out maintenance offshore means that offshore wind turbines need to be extremely reliable. Most are now equipped with real-time condition monitoring systems that can highlight the development of potential faults. Systems that allow modification of turbine functions to work around the fault until it can be rectified are also becoming important in offshore wind turbine operations.

In order that they can be maintained, offshore wind turbines must have facilities to allow support vessels to moor and transfer staff. The turbines may also be able to receive maintenance staff by helicopter. Whether maintenance crews arrive by sea or air, this still means that the turbines must be serviced from a shore base. As offshore development advances, particularly where wind farms are located far from shore, it may become practical to build a "mother platform," similar to an oil and gas platform, alongside a wind farm to house the wind farm substation and to provide accommodation for operations and maintenance staff.

FOUNDATIONS FOR OFFSHORE WIND TURBINES

While the technology for offshore wind turbines is broadly similar to that used onshore, there is one area where the two types differ significantly and that is in the nature of their foundations. While onshore wind turbines are usually anchored using relatively conventional gravity foundations, offshore wind turbines normally require entirely different support structures. These can account for as much as 30% of the capital cost of the wind turbine installation.

The type of foundation that is used depends on the offshore location, the water depth, and the seabed geology. Several types are shown in Figure 9.1. In shallow waters close to the shore it is possible to use a type of gravity foundation. These can be used up to depths of 20 m or more

Figure 9.1 Foundations for offshore wind turbines. Source: NREL (Large-Scale Offshore Wind Power in the United States ASSESSMENT OF OPPORTUNITIES AND BARRIERS September 2010 NREL/TP-500-40745).

but will usually be limited to much shallower waters. More common is the monopole, or monopile, a single pile driven into the seabed onto which the turbine tower is bolted. Beyond around 30–40 m these too become difficult to establish and more complex seabed structures are necessary. For depths of 60 m and greater, all seabed foundations become impractical and some form of floating support will be necessary.

The gravity foundation is a massive structure that sits on the seabed. It must be massive enough to withstand the bending and translational forces that are transferred to it through the turbine tower. For shallow offshore sites, concrete caisson-type structures can be fabricated onshore and then floated to the turbine or wind farm site where they are filled with rock ballast and sunk. With this type of foundation it is important that the seabed where each caisson is sunk should be completely level, otherwise the foundation will not sit squarely and the tower will not be vertical.

An alternative to the gravity foundation for shallow waters, though rarely used in practice, is the suction bucket. This uses an upturned vessel that is placed on the seabed and then the water and air inside it is evacuated, causing the external weight of water and atmospheric

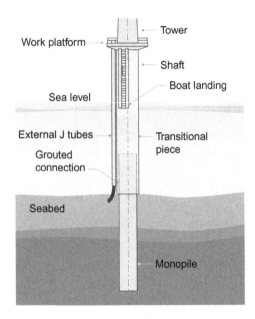

Figure 9.2 A monopile foundation for an offshore wind turbine. Source: Courtesy of E.ON Energy http://www. eon-uk.com/downloads/Foundation_types_Monopile.pdf.

pressure to force it into the seabed. This type of foundation can only be used when the seabed is relatively soft.

The most commonly used offshore foundation is the monopole or monopile foundation. As its name suggests, this is a single pile that is sunk into the seabed. The pile must reach through any silt on the seabed to a stable stratum below, so its depth below the seabed will depend on the geology of the area. The pile foundations are created using special vessels. Figure 9.2 shows a monopile foundation in detail.

The depth of the monopile above the seabed is typically 30–50 m with most of this underwater. The upper end is provided with a flange and the actual wind turbine tower is then erected from this flange, in sections, as would be an onshore wind turbine tower. Once the tower has been built, the nacelle structure must then be raised onto it. This can be carried out using large lifting vessels that can lift the several hundred tonnes to the top of the tower which may be 100 m or more above sea level. An alternative is to build in a "jack-up" system whereby the nacelle and rotor are hoisted to the top using the tower, once it has been erected.

Monopile foundations extend the length of the tower compared to an onshore wind turbine. An offshore wind turbine with a 50 m monopile, to which a 100 m tower is bolted will have a single support of 150 m in length. This can aggravate the problems of tower resonant frequencies and has to be taken into account when designing offshore installation using this type of foundation. It will also be one of the factors limiting the maximum length of the monopile foundation. Building adequate stiffness into a monopile/tower combination in deep waters will be extremely costly in materials.

When the depth of water at an offshore wind farm site becomes too large—usually somewhere between 30 and 50 m, the monopile is no longer the best solution. Instead a structure with several legs will normally be preferred. These include multipile structures, tripods, and space frames. A multipile foundation has the tower support structure itself supported on, typically, three small piles instead of one large one. This will improve stability at deep water sites. Tripods and spaceframes are similar in concept but instead of being piled into the seabed they are anchored, with the legs providing the necessary stability.

Deep water caisson installations are also being studied. These can be anchored simply using their mass or some form of suction installation is possible.

FLOATING WIND TURBINE FOUNDATIONS

The necessity to create seabed foundations for offshore wind turbines significantly limits the areas of sea accessible to offshore wind. When sea depths extend much beyond 60 m, it becomes difficult to create cost-effective foundations of this type. As a consequence there has been extensive research in the past decade into the practicability of using floating supports for wind turbines instead. These could potentially revolutionize the costs of offshore wind because the supports could be put together onshore and simply floated or carried by ship to the wind farm site. All that would then be required is a suitable anchoring system.

There are five or six prototype floating platforms for wind turbines in operation around the world, half of them in Japan. Designs vary as different concepts are tested. Most developments are aimed at building an individual platform for each wind turbine, though some designs could carry more than one. These are often complex semisubmersible structures which attain stability by, like an iceberg, having most of the structure underwater. Simpler spar buoy supports are also being tried. All are anchored to the seabed with multiple stays to help maintain stability and position.

There are some advantages to using vertical axis wind turbines on floating platforms because all of the drive train components can be located at the platform level and only the rotor has to be lifted high into the air. The disadvantage is that unless a vertical axis rotor is placed at the same height as a typical horizontal axis wind turbine it cannot take advantage of the best wind available. Tall vertical rotors capture much less energy near the ground than they do at the top of the rotor. So while vertical axis systems are being explored, most development is focused on horizontal axis turbines.

The advantage of floating platforms, apart from cost, is that they allow turbines to operate in deep waters, often far offshore, where the wind blows more strongly and more regularly, enabling the machines to generate more power. The US Department of Energy believes that floating wind turbines offer the best option for the United States where 60% of the nation's offshore wind resources are in deep water. Japan too has extensive offshore wind resources in deep water regions.

OFFSHORE WIND TURBINE SIZE

Whatever the choice of foundation or support for an offshore wind turbine, the capital cost will always be higher than for a similarly sized onshore turbine. This has forced developers to focus on the economics of offshore wind farms and particularly on wind turbine size. If the support structure is costly, then it makes sense to install the largest turbine possible onto it. In addition, the better wind regime offshore means that the returns for increasing wind turbine size are greater than they would be onshore.

In 2014 the average size of all offshore wind turbines was 4 MW, significantly larger than the average for onshore machines. Meanwhile wind turbines for offshore use are becoming larger and larger with the average size of commercial machines around 5 MW in 2015. Major manufacturers are developing much larger machines still. Commercial wind turbines of between 7 and 8 MW are already available and companies are designing and testing even larger turbines with capacities of 10−15 MW.

Wind turbines of this size would be extremely difficult to build onshore using conventional construction methods because the size of the blades and tower sections are so large that they cannot easily be transported by road. Such restrictions do not exist at sea. Although the marine environment presents difficulties of its own, wind turbine size is not restricted in the same way. It is not clear if there is a practical limit to the size of an offshore wind turbine but long term projections include plans for 20 MW machines by the end of the second decade of the twenty-first century.

INFRASTRUCTURE CONSIDERATIONS AND OFFSHORE GRIDS

Construction of offshore wind farms has become a highly specialized activity and the industry is beginning to concentrate in particular areas. The difficulties associated with transporting large wind turbines on land means that the ideal site for a manufacturing facility is at the coast, as close as possible to the area in which wind turbines are being installed, the east coast of the United Kingdom for North Sea wind farms, for example. This will eventually lead to the creation of offshore wind turbine industrial clusters where manufacturers, transportation vessels, and the specialist foundation making and turbine erection vessels all operate from within a small region. Maintenance companies will probably operate from the same cluster and, as noted above, it may eventually be practical to built maintenance facilities on a platform within a wind farm once developments become sufficiently far from land.

One of the other key considerations is how to bring the power from offshore wind farms to the shore and to the main demand centers. Onshore wind farms will connect to the nearest section of the distribution or transmission system but there are no transmission and distribution systems offshore so new connections must be established with each wind farm.

Most wind farms today are connected to the shore on a farm-by-farm basis. Depending on the distance from the shore this connection will be a high voltage AC link, or a high voltage DC (HVDC) interconnection. For distances from wind farm to the shore of up to 100 km the AC connection is probably the most economical way of delivering power to the shore. However, there is a cable-to-sea capacitance associated with each part of the undersea cable which increases with cable length and requires reactive power compensation. Beyond around 80 km, therefore, it becomes economically practical to consider an HVDC link instead. Transmission capacity of the two types of interconnection is a factor too, with the HVDC link being more economical for transmitting large volumes of power.

The farm-by-farm connection scheme becomes more of a problem as wind farms are located further and further from the shore. An alternative to this approach is to build an offshore grid that can be used to interconnect all the wind farms in an area and then bring the power ashore from the grid at several interconnection points. An offshore grid of this type has been proposed for the North Sea. As well as interconnecting wind farms it would connect to several nations around the North Sea, providing a much strengthened grid system between the countries. A similar grid has also been proposed for the western coast of Europe, stretching from the North Sea to the Portuguese coast.

An offshore grid would probably be built using HVDC technology since this appears more effective for underwater use. The technology for multiple node HVDC grids is in an early stage of development but appears practical. It is expected to be based on transistor-based voltage source conversion (VSC) technology rather than the more traditional thyristor-based technology that has been widely used for HVDV interconnections. VSC technology has already been used to connect one European wind farm to the shore.

CHAPTER 10

Wind Power Grid Integration and Environmental Issues

The rapid growth of wind power has presented grid operators with a number of problems related to its integration into the electricity supply system. The intermittent and unpredictable nature of wind means that it is more difficult to plan wind power grid inputs on a daily or weekly basis. In addition, where there is a significant volume of wind generating capacity connected to a grid, the operators must also have alternative capacity available to step in when the wind fails. Managing this represents one of the greatest challenges for wind power.

There are other challenges. The best areas for building wind farms and erecting wind turbines are often a long way from the places where the bulk of the electricity is needed. While some wind developments can serve local demand, the construction of large blocks of wind power in the best wind sites—in Texas or California or offshore in the North Sea, for example—means that the power available is much more than any local demand could consume. Bringing the wind energy to demand centers means building new transmission lines but it may also mean dramatically reconfiguring existing transmission networks in order to make it possible to carry the power to where it is required.

The development of wind energy also throws up environmental issues. Though many of these have been addressed over the past 35 years, environmental considerations will normally form the part of any modern wind energy development.

WIND INTEGRATION AND GRID OPERATION

The modern electricity grid has been designed to supply consumers with power and that power has come traditionally from central power stations. These power stations are divided into different categories depending upon the cost of the electricity they produce and the ease with which their output can be varied. There are base load plants,

Wind Power Generation. DOI: http://dx.doi.org/10.1016/B978-0-12-804038-6.00010-4

normally the cheapest source of electricity on the network, that operate flat out virtually all the time. Intermediate load plants will be used to meet the increased demand during daylight hours but they may shut down overnight. Finally peak load plants only supply power intermittently when there is a short term surge in demand for electricity. In this way, the grid operator can provide exactly the right amount of electrical power to balance the level of demand at any time, day or night.

In this conventional grid there is always variability and supply levels are always changing, but it is only the changes in demand that the operator must match. When wind power is added to the grid it becomes not only demand that varies, but supply that becomes variable too. In most cases wind power will be supplied to a grid automatically if it is available. However, the amount will depend on the strength of the wind. The operator must now match the variations in wind energy supply with changes in the supply from more conventional sources as well as matching the varying level of demand.

There are a number of strategies that can be applied to assist in managing this new variable source of energy. The first is good weather forecasting. Modern weather predicting is now extremely good and can provide a reliable picture of the wind across a network region. Weather forecasting data is available from a number of sources today and computer-based wind energy forecasting systems, which are often built into the grid operator's management system, will probably use data from several sources to create the most reliable forecast possible.

There are two approaches to wind energy forecasting for grid system operators, decentralized and centralized. With decentralized forecasting a wind energy output forecast is generated for each wind turbine or wind farm that is connected to the grid. Centralized forecasting, in contrast, provides a network-region wide prediction of the amount of wind energy that is likely to be available. Using forecasts of this type which might predict the wind energy for the next day, or for the next hour, the system operator can plan to have alternative sources of electricity ready to come on line to match the forecast changes in wind energy input.

Those alternative sources of electricity form another key part of wind energy management and integration. If the amount of wind energy being provided to a grid is varying on an almost continuous basis, the grid must have other generating units that can either cut back their output as wind input rises or take up the slack if the wind input suddenly falls. In the past peak demand has typically been supplied by fast acting simple gas turbine power plants but these are expensive to run. In order to provide support for wind energy, many grids instead use combined cycle power plants that are cheaper to run. These plants would not traditionally have supplied grid backup of this type and they often have to be adapted to operate successfully in this way. Coal-fired plants may also be able to support wind energy although they are more normally considered to be base load power plants and cannot change usually their output levels very quickly. Hydropower plants, especially if they are power plants with dams and storage reservoirs can also be extremely useful in helping to integrate wind power. These plants are fast acting and so can respond very quickly to changes in the supply level. In countries or regions where hydropower is abundant, wind energy integration has proved much cheaper to achieve than elsewhere.

ENERGY STORAGE

Another valuable way of backing up wind energy on a grid is by using energy storage plants. These are also capable of responding very quickly to changes in demand and some are ideally suited to wind energy support. Technologies such as pumped-storage hydropower, compressed air energy storage, and batteries can be used in this way.

Energy storage can be operated on a grid-wide basis, but it can also be operated on a local basis with a single storage facility dedicated to a single wind farm. The principle of operation is that when there is surplus energy available on the grid, it is stored in the energy storage facility. Then, when demand picks up the stored energy can be used to supply the additional demand.

At a local level, off-grid wind turbines are often used to provide power to charge batteries. Depending upon the size of the wind turbine and the capacity of the batteries, this type of system is capable of providing continuous power to a consumer. Very large batteries can also be used on the grid as a way of storing power.

Individual wind farms can also be built with their own energy storage capacity. Again, by sizing the battery and wind farm generating capacities correctly, it is possible for a single wind farm to be able to provide power on demand, rather than only when the wind blows. This allows it to perform a more valuable role of grid support which, in turn, is likely to earn the wind farm operator greater revenue.

For grid-level energy storage, the most successful technology is pumped storage hydropower. This type of power plant has two reservoirs, one high and one low, and a hydroturbine that can also act as a pump. When the plant needs to generate electricity, water is taken from the high reservoir and passed through the turbine, then collected in the lower reservoir. To store energy, surplus electricity is used to pump water from the lower to the upper reservoir. Plants of this type can have capacities of several thousand of megawatts.

A more unusual way of storing wind energy is to use the surplus power from a wind farm to make hydrogen from water by electrolysis. Hydrogen can be stored and then either used to produce more electricity, or as fuel for a hydrogen powered vehicle. Hydrogen can also be transported easily and could provide an alternative to natural gas in a carbon-free energy system. While no commercial projects exploit this idea there are a number of test projects under way.

GEOGRAPHICAL AVERAGING OF WIND OUTPUT

Instead of trying to manage the variable nature of wind power with alternative sources, a different approach is to try and make the power from wind more reliable. One way of doing this is by using geographical averaging. A single wind turbine will be subject to all the vicissitudes of the local wind. Sometimes it will blow and sometimes it will not. However when the wind is not blowing at this particular wind energy site it will often be blowing at another. If the two sites are close together then they probably share similar weather. The farther apart they are, the more different their wind regimes are likely to be.

Geographical averaging involves treating these two wind turbines as a single power plant. It is more likely that one or the other of them will be generating power at a specified time than either one of them alone. And if this is extended to include many wind turbines, and if they are spread over a large geographical area, then the output from them, taken as a whole becomes much more reliable.

Of course, there will still be significant output variations but even so, the reliability of this combined wind resource will be greater. Such averaging can be carried out by grid operators but from a commercial point of view it may be more effective for the plant owners to combine wind plants into a "virtual wind farm." This will allow a group of wind farms to appear to the grid as a single wind power plant. In doing so, the owners can increase the value of the energy they supply because from the grid point of view it is more predictable.

WIND INTEGRATION LIMITS

How much wind power can a single grid support before it becomes unstable? This is an important question but answering it is not easy. Clearly wind power cannot provide 100% of grid power because there can be extended periods when the wind does not blow. Though rare, if one of these calms fell over a region, even if it was equipped with massive energy storage capacity, the power would eventually run out.

On the other hand, there is already evidence from countries such as Germany and Denmark that relatively high levels of wind energy are supportable. In both of these countries there are examples of regions that have operated with 30% of all electric power coming from the wind.

The ability of any grid to support large volumes of wind will depend on its structure and on the energy plants connected to it. An analysis by the International Energy Agency in 2011 indicated that the Danish grid should be able to support 63% of its energy coming from variable renewable sources, mainly wind. For the Nordic market as a whole, the estimate was 48%. At the other end of the scale, Japan was judged only to be able to support 19%. Much depends on the nature of the grid in each country or region and the amount of flexible power generation there is available.

WIND ENERGY AND GRID STRUCTURAL ISSUES

The structural issues affecting wind integration into a grid are determined by the layout of a transmission grid, and to a lesser extent the distribution grid. The type of problem that can arise is exemplified by the experience of Germany during the twenty-first century. Here wind energy has been strongly promoted as a replacement for fossil fuel and nuclear power. However most of the best wind resources are in the north of the country and in the offshore waters of the northern coast. Meanwhile most of the major demand centers where the power is needed are in the south. Germany's grid system was designed around large coal-fired and nuclear power plants feeding power to these demand centers. These plants were not generally located in northern Germany and so the grid was not designed to transport large volumes of power from the north to the south. Having to do so as wind generation has expanded has exposed this inability of the German grid to handle the new power flows. The solution is to modify and reinforce the grid but in a highly developed country like Germany it is not easy to build new transmission lines. The problem is therefore taking a long time to solve.

Similar situations have arisen in the United States where the best wind zones are in the Midwest, in Texas, and to the south west, while the greatest demand for power is in the northeast. Reconfiguring the grid with new transmission lines is taking place here too but planning issues in the United States are not quite so severe as in Germany. Even so, adaptation takes time.

There is an even bigger reconfiguration problem to solve in Europe. The European Union is trying to establish an open market for electricity. In principle this would allow wind power from Germany and the Atlantic coast regions to be transported all over Europe, as well as solar power from countries in the south of the continent. However the national grids in the individual countries are currently incapable of handling the power flows that would be required and there are significant bottlenecks between countries.

Meanwhile in China the massive growth of wind power, particularly in the northern regions of the country is forcing the construction of new transmission corridors to carry the power to the locations where it is needed. As elsewhere, solving the problem takes time and is costly.

WIND POWER AND THE ENVIRONMENT

The main environmental issue that all wind turbine and wind farms developers must face is visibility. A modern utility wind turbine is a large machine and it cannot be hidden from view. Some people consider them beautiful. Others think they are an offence to the eye. Since both are opinions are subjective, there is no way of judging either to be correct.

In some countries, such as the United Kingdom, it has become extremely difficult to obtain the necessary permission to build onshore wind farms because of objections to the visual intrusion they cause. In other countries there are sufficient remote sites that has is not become such a limiting issue. Building offshore is easier because offshore wind farms have much less of a visible impact. If they are built far enough from the shore they become completely invisible.

Other problems associated with wind turbines include noise, damage to wildlife, and interference with various electrical communication and radar systems. Modern utility turbines turn relatively slowly. However, they do create wind noise related to the speed at which the blade tips pass through the air. This speed is usually kept below a maximum level to limit the aerodynamic noise generated by the blades as they rotate. The machinery within the nacelle will also generate mechanical noise. Some of this can be damped with good nacelle design. However, turbines cannot usually be sited close to habitations because of the noise generated.

In the past wind turbines have been blamed for bird deaths. This has become less of a problem with slower moving blades although "bird strikes" will still occur. Offshore there may also be effects on marine life although these are generally judged to be slight, except during the construction period for an offshore wind farm when there will be considerable seabed disruption.

Wind turbines can interfere with radar systems and so affect aircraft and airport safety. Modern technology can usually overcome these problems although in some cases projects have been halted for this reason. Often the problem lies with aging radar systems but even here relatively simple operational changes can often overcome the issues. Radio interference is also possible but will usually only be found very close to a turbine.

CHAPTER 11

The Cost of Electricity from Wind Turbines

The cost of electricity from a power plant of any type depends on a range of factors. First there is the cost of building the power station and buying all the components needed in its construction. In addition, most large power projects today are financed using loans so there will also be a cost associated with paying back the loan, with interest. Then there is the cost of operating and maintaining the plant over its lifetime. Finally the overall cost equation should include the cost of decommissioning the power station once it is removed from service.

It would be possible to add up all these cost elements to provide a total cost of building and running the power station over its lifetime, including the cost of decommissioning, and then dividing this total by the total number of units of electricity that the power station produced over its lifetime. The result would be the real lifetime cost of electricity from the plant. Unfortunately, such a calculation could only be completed once the power station was no longer in service. From a practical point of view, this would not be of much use. The point in time at which the cost-of-electricity calculation of this type is most needed is before the power station is built. This is when a decision is made to build a particular type of power plant, based normally on the technology that will offer the least cost electricity over its lifetime.

LEVELIZED COST OF ENERGY MODEL

In order to get around this problem economists have devised a model that provides an estimate of the lifetime cost of electricity before the station is built. Of course, since the plant does not yet exist, the model requires a large number of assumptions and estimates to be made. In order to make this model as useful as possible, all future costs are also converted to the equivalent cost today by using a parameter known as

Wind Power Generation. DOI: http://dx.doi.org/10.1016/B978-0-12-804038-6.00011-6

the discount rate. The discount rate is almost the same as the interest rate and relates to the way in which the value of one unit of currency falls (most usually, but it could rise) in the future. This allows, for example, the maintenance cost of a wind turbine 20 years into the future to be converted into an equivalent cost today. The discount rate can also be applied to the cost of electricity from the wind power plant in 20 years time.

The economic model is called the levelized cost of electricity (LCOE) model. It contains a lot of assumptions and flaws but it is the most commonly used method available for estimating the cost of electricity from a new power plant.

When considering the cost of new power plants the levelized cost is one factor to consider. Another is the overall capital cost of building the generating facility. This has a significant effect on the cost of electricity but it is also important because it shows the financial investment that will have to be made before the power plant generates any electricity. The comparative size of the investment needed to build different types of power stations may determine the actual type of plant built, even before the cost of electricity is taken into account. Capital cost is usually expressed in terms of the cost per kilowatt of generating capacity to allow comparisons between technologies to me made.

When comparing different types of power station there are other factors that need to be considered too. The type of fuel, if any, that is used is one. A coal-fired power station costs much more to build than a gas-fired power station but the fuel it burns is relatively cheap. Its price rarely changes dramatically either. Natural gas, is more expensive than coal and it has historically shown much greater price volatility than coal. This means that while the gas-fired station may require lower initial investment, it might prove more expensive to operate in the future if gas prices rise dramatically.

Renewable power plants can also be relatively expensive to build. However they normally have no fuel costs because the energy they exploit is from a river, from the wind, or from the sun and there is no economic cost for taking that energy. That means that once the renewable power plant has been paid for, the electricity it produces will have a very low cost. All these factors may need to be balanced when making a decision to build a new power station.

WIND TURBINE CAPITAL COSTS

The capital cost of a wind power plant will be made up of several components. The largest is the cost of the wind turbine itself, normally followed by the cost of building a foundation. For offshore wind plants this can prove almost as costly as the turbine. Then there is the cost of transporting the turbine to the wind site, a figure that can be substantial as the size of turbines increases and finally the cost of connecting the turbine to the wind farm substation (assuming the unit is part of a wind farm) and connecting the substation to the grid.

Turbine costs will depend on global commodity costs. For example the towers for wind turbines require significant amounts of iron. This limited the ability of developers to build towers much above 80 m during the latter years of the first decade of the twenty-first century because of the elevated cost of iron on the global market. The cost of copper—for transformer and generator windings and for the interconnections—or carbon fiber if this is used in blades, all affect the final turbine cost.

Table 11.1 shows the overnight cost[1] of wind turbines in the United States between 2001 and 2014 from figures published each year by the US Energy Information Administration in its Annual Energy Outlooks. The figures in the table show onshore wind turbine costs for the whole period. However, offshore wind energy costs were only included in the reports towards the end of the period shown in the table.

Table 11.1 The Capital Cost of Wind Power Plants in the United States		
Year	Overnight Cost of Onshore Wind ($/kW)	Overnight Cost of Offshore Wind ($/kW)
2001	919	–
2003	938	–
2005	1060	–
2007	1127	–
2009	1797	–
2011	2251	4404
2013	2032	4452
2014	2061	4503
Source: US Energy Information Administration Annual Energy Outlooks 2001–2014.		

[1]The overnight cost is the cost of the wind turbine without any loan repayments so it simply represents the cost of making the turbine at that point in time.

The figures show a steady increase in the cost of onshore wind power between 2001 when the overnight cost was $919/kW to 2007 when it had risen to $1127/kW. However, there was a step change in the cost between 2007 and 2009 when the estimated overnight cost was $1797/kW, close to 60% higher. This is probably associated with a rapid rise in global commodity prices during this period. The price rose again, to $2251/kW in 2011 before dropping back to $2032/kW in 2013; again this is probably mostly due to a change in commodity prices although competition between wind turbine manufacturers was becoming more intense during this period too. In 2014 the overnight cost was estimated to be $2061/kW, a modest price rise.

Offshore wind capital costs are much higher than those for onshore wind due to the greater difficulty in installing wind turbines offshore. In 2011, the first year for which offshore figures are quoted in the table, the overnight cost was put at $4404/kW. This had risen to $4503/kW by 2014. These figures are for the United States and they can be expected to vary from region to region. However, as the cost of wind turbines become global, variations in capital costs from one country to another will become smaller.

THE LEVELIZED COST OF WIND ENERGY

For wind power, the main component that determines the levelized cost of electricity is the capital cost of buying and installing the wind turbine(s) and connecting it to the grid. Operational and maintenance costs are the other key factor.

Table 11.2 contains figures, again for the United States for the levelized cost of wind power, based on modeling carried out by Lazard and published in 2014. The table shows a range of onshore wind power

Table 11.2 The Levelized Cost of Electricity from Wind Power Plants in the United States	
	Levelized Cost of Electricity ($/MWh)
Onshore wind power	37–81
Offshore wind power	162
Source: *Lazard's levelized cost of energy analysis—Version 8.0, Lazard 2014.*	

costs, from \$37/MWh to \$81/MWh. The lowest cost is found in the US Midwest, while the highest wind energy cost is found in the southeast of the United States. Lazard has found a 58% decrease in its calculated levelized cost of wind power between 2009 and 2014.

The cost of offshore wind power in the United States is estimated to be \$162/MWh, significantly higher than for onshore wind. However, like onshore wind energy costs, this will vary from region to region. Nevertheless, offshore wind is likely to remain more expensive than onshore wind unless a radical change to offshore deployment, such as the use of floating platforms, becomes commercially viable.

Printed in the United States
By Bookmasters